Other Monographs in this Series

Conference Board of the Mathematical Sciences

REGIONAL CONFERENCE SERIES IN MATHEMATICS

supported by the

National Science Foundation

Number 16

MEASURE ALGEBRAS

by

JOSEPH L. TAYLOR

Published for the

Conference Board of the Mathematical Sciences

by the

American Mathematical Society

Providence, Rhode Island

Expository Lectures
from the CBMS Regional Conference
held at the University of Montana
June 1972

AMS 1970 Subject Classifications: 43-02, 43A10, 43A20; 46J20, 46J25

Library of Congress Cataloging in Publication Data

Taylor, Joseph L 1941–
Measure algebras.

(Regional conference series in mathematics, no. 16)
Bibliography: p.
1. Measure algebras. 2. Banach algebras.
3. Semigroups. I. Conference Board of the Mathematical Sciences. II. Title. III. Series.
QA1.R33 no. 16 [QA403] 510′.8s [512′.55]
ISBN 0-8218-1666-7 73-5930

Preface

These notes were prepared in conjunction with the N. S. F. regional conference on measure algebras held at the University of Montana during the week of June 19, 1972.

Our original objective in preparing these notes was to give a coherent detailed and simplified presentation of a body of material on measure algebras developed in a recent series of papers by the author (Taylor [1] – [10]). This material has two main thrusts: the first concerns an abstract characterization of Banach algebras which arise as algebras of measures under convolution (convolution measure algebras) and a semigroup representation of the spectrum (maximal ideal space) of such an algebra; the second deals with a characterization of the cohomology of the spectrum of a measure algebra and applications of this characterization to the study of idempotents, logarithms, and invertible elements.

As this project progressed the original concept broadened. The final product is a more general treatment of measure algebras, although it is still heavily slanted in the direction of our own work.

Chapter 1 contains a brief introductory discussion of convolution and the structure of the algebras $L^1(G)$ and $M(G)$, as well as an introduction to several of the problems which will be solved or partially solved in later chapters.

Chapters 2 and 3 are devoted to a development and discussion of convolution measure algebras and to a representation theorem for the spectrum of such an algebra. Several examples of convolution measure algebras are discussed in Chapter 4. Much of the material of Chapters 2–4 is contained in Taylor [1] and can be skipped by readers familiar with that paper. However, our discussion here is considerably more detailed and does not assume familiarity with Kakutani's L-space theory or the theory of topological semigroups.

Chapters 5–9 are mainly concerned with a characterization of the cohomology of the spectrum of a measure algebra and applications to the study of idempotents, logarithms, and inverses in such an algebra. This material originally appeared in Taylor [3] – [10]. The development here has been considerably simplified.

Chapter 10 is largely independent of Chapters 4–9. In it we discuss some results of Miller [1] on Gleason parts in a measure algebra, of Taylor [2] and Johnson [3] on the Shilov boundary of $M(G)$, and of Brown and Moran [3] on infinite product measures.

We would like to express our gratitude to Michael J. Fisher who conceived and organized the conference, and to the National Science Foundation, the University of Montana, the Rocky Mountain Mathematics Consortium, and the Conference Board of the Mathematical Sciences. We also acknowledge the generous financial support of the Air Force Office of Scientific Research under grant No. 1313-67, the National Science Foundation under grant No. GP-32331, and the Alfred P. Sloan Foundation.

University of Utah

Joseph L. Taylor

TABLE OF CONTENTS

CHAPTER 1. ORIENTATION

We shall assume that the reader is familiar with elementary commutative Banach algebra theory, including the definitions and elementary properties of the spectrum (maximal ideal space) and Gelfand transform. Our main source for background in this area will be Gamelin [1]. In addition, we shall assume some familiarity with abstract harmonic analysis as presented in Rudin [5].

In this chapter we present several examples and problems related to measure algebras, in order to motivate the reader for what will follow.

§1.1. Measure algebras on semigroups

1.1.1. A topological semigroup is a Hausdorff topological space S with an associative operation $(s, t) \to st : S \times S \to S$ which is jointly continuous.

1.1.2. CONVOLUTION. We denote the Banach space of complex, finite, regular Borel measures on a locally compact Hausdorff space X by $M(X)$.

If S is a locally compact topological semigroup, then $M(S)$ is a Banach algebra under the operation $(\mu, \nu) \to \mu * \nu$ (convolution), defined by the condition

(i) $$\int f \, d\mu * \nu = \iint f(st) \, d\mu(s) \, d\nu(t)$$

for $f \in C_0(S)$. Recall that $M(S)$ is the dual space of $C_0(S)$, the space of continuous functions vanishing at infinity on S; hence, (i) defines a measure $\mu * \nu \in M(S)$ for each pair $\mu, \nu \in M(S)$.

It is easy to see that convolution is associative and distributive and satisfies $\|\mu * \nu\| \leqslant \|\mu\| \, \|\nu\|$. Hence, $M(S)$ is a Banach algebra under convolution. Furthermore, $M(S)$ is commutative if and only if S is abelian; and if S has an identity e, then the point mass δ_e at e is an identity for $M(S)$.

It follows from standard facts of integration theory that (i) holds for all bounded Borel functions f if it holds for functions in $C_0(S)$. In particular, with $f = \chi_E$, the characteristic function of E, we have

(ii) $$\mu * \nu(E) = \iint \chi_E(st) \, d\mu(s) \, d\nu(t).$$

1.1.3. L. C. A. GROUPS. If a topological semigroup S has an identity e and a continuous inverse map $s \to s^{-1} : S \to S$ with $ss^{-1} = s^{-1}s = e$, then S is called a topological group.

Of particular interest to us will be locally compact abelian topological groups (l. c. a.

groups). If G is an l.c.a. group then the measure algebra $M(G)$ is commutative and has an identity δ_e (which we shall ordinarily denote simply by δ). Furthermore, the inversion map $g \to g^{-1}$ induces an involution $\mu \to \tilde{\mu}$ on $M(G)$ defined by $\tilde{\mu}(E) = \overline{\mu(E^{-1})}$.

1.1.4. GROUP ALGEBRAS. If G is an l.c.a. group, then there is a unique (up to a multiplicative constant) positive translation invariant measure m on G. This is the Haar measure of G (Naimark [1]).

We denote the subspace of $M(G)$ consisting of all finite measures absolutely continuous with respect to Haar measure m by $L(G)$. Note that $L(G)$ is a closed subspace of $M(G)$. Furthermore, since the collection of null sets for m is translation invariant, we have that

$$\mu * \nu(E) = \iint \chi_E(g_1 g_2)\, d\mu(g_1)\, d\nu(g_2) = \int \mu(g_2^{-1}E)\, d\nu(g_2) = 0$$

whenever $\mu \in L(G)$, $\nu \in M(G)$, and E is a null set for m. It follows that $L(G)$ is an ideal of $M(G)$.

Now if $f \in L^1(G) = L^1(m)$, then $fm \in L(G)$, where $fm(E) = \int \chi_E f\, dm$. In fact, $\|f\|_1 = \|fm\|$ and, by the Radon-Nikodym theorem, every measure in $L(G)$ has this form. Hence, $L^1(G)$ and $L(G)$ are isometric Banach spaces under the map $f \to fm$. This map becomes an algebraic isomorphism if we define convolution for $f, h \in L^1(G)$ by

$$f * h(g) = \int f(g_1) h(g_1^{-1} g)\, dm(g_1).$$

1.1.5. SEMICHARACTERS. If S is a semigroup, then a semicharacter on S is a homomorphism of S into the multiplicative complex plane, that is, a complex valued function f on S such that $f(st) = f(s)f(t)$ for $s, t \in S$.

If S is a topological semigroup, then $\hat{S}$ will denote the space of all continuous, nonzero, bounded semicharacters on S. Note the fact that $f(s^n) = f(s)^n$ for $f \in \hat{S}$, $s \in S$ implies that $|f| \leq 1$ for each $f \in \hat{S}$.

If S is a locally compact abelian semigroup, then each $f \in \hat{S}$ determines a complex homomorphism $\mu \to (f, \mu)$ of $M(S)$ by $(f, \mu) = \int f\, d\mu$; in fact,

$$\begin{aligned}(f, \mu * \nu) &= \iint f(st)\, d\mu(s)\, d\nu(t) = \iint f(s)f(t)\, d\mu(s)\, d\nu(t) \\ &= (f, \mu)\,(f, \nu)\end{aligned}$$

for $f \in \hat{S}$, $\mu, \nu \in M(S)$. Thus, $\hat{S}$ determines at least part of the spectrum of $M(S)$. However, in general there are complex homomorphisms in the spectrum of $M(S)$ which do not arise this way.

1.1.6. THE SPECTRUM OF $L(G)$. On a l.c.a. group G, each semicharacter f must have modulus one, since $|f(g)| \leq 1$ for all $g \in G$ and $f(g^{-1}) = f(g)^{-1}$. The semicharacters on a group are called characters. The space $\hat{G}$ of all continuous characters is a group under multiplication and is called the dual group of G.

While it is not true that every complex homomorphism of $M(G)$ is determined by an element of $\hat{G}$, for the subalgebra $L(G)$ this is true. In fact, each complex homomorphism of $L(G)$ has the form $\mu \to \int f\, d\mu$ for some $f \in L^\infty(\mu)$ satisfying

$$\text{(i)} \quad \begin{aligned}\iint f(g_1g_2)\,d\mu(g_1)\,d\nu(g_2) &= \int f\,d\mu*\nu = \int f\,d\mu\int g\,d\nu \\ &= \iint f(g_1)f(g_2)\,d\mu(g_1)\,d\nu(g_2) \quad \text{for } \mu,\nu\in L(G).\end{aligned}$$

Now the function $\phi(g) = \int f(g_1g)\,d\mu(g_1)$ is continuous for each $\mu \in L(G)$ and satisfies $\phi(g) = f(g)\phi(0)$ a. e. by (i).

Since μ can be chosen so that $\phi(0) = \int f(g_1)\,d\mu(g_1) \neq 0$, we conclude that f is equal a. e. to a continuous function $\gamma(g) = \phi(g)\phi(0)^{-1}$. It then follows from (i) that $\gamma(g_1g_2) = \gamma(g_1)\gamma(g_2)$ and, hence, $\gamma \in \hat{G}$.

Thus, we may identify the spectrum of $L(G)$ with the space $\hat{G}$, and the Gelfand transform for $L(G)$ with the Fourier transform $\mu \to \hat{\mu}$, where

$$\text{(ii)} \qquad \hat{\mu}(\gamma) = \int \gamma(g)\,d\mu(g) \quad \text{for } \gamma \in \hat{G}.$$

It turns out that $\hat{G}$ is an l. c. a. group with the Gelfand topology induced by $L(G)$. Furthermore, the natural embedding of G in $\hat{\hat{G}}$ defined by $g \to \tilde{g}$ $(\tilde{g}(\gamma) = \gamma(g))$ is an isomorphism and homeomorphism of G onto $\hat{\hat{G}}$ (cf. Rudin [5]).

§1.2. **Convolution equations**

1.2.1. Given $k \in L^1(R)$, $\lambda \in \mathbf{C}$, the convolution integral equation determined by k and λ is the equation

$$\text{(i)} \qquad \lambda f(x) + \int f(y)k(x-y)\,dy = h(x).$$

For a given p $(1 \leqslant p \leqslant \infty)$ we consider the question of when this equation has a unique solution $f \in L^p(R)$ for each $h \in L^p(R)$.

If we define $\mu \in L(R) + \mathbf{C}\delta \subset M(R)$ by $\mu = \lambda\delta + km$ (m Lebesgue measure), then this equation becomes

$$\text{(ii)} \qquad C_\mu f = \mu * f = h,$$

where $\mu * f(x) = \int f(x-y)\,d\mu(y)$ and the first equality in (ii) defines the operator $C_\mu : L^p(R) \to L^p(R)$.

Equation (ii) has a unique solution for each $h \in L^p(R)$ if and only if the operator C_μ is invertible. It is well known that for each p, C_μ is invertible if and only if the Fourier transform $\hat{\mu} = \lambda + \hat{k}$ is nonvanishing on $R \cup \{\infty\}$. As we shall see, this is a consequence of the fact that $\hat{R} = R$ is the spectrum of $L(R)$.

1.2.2. Convolution Operators. If G is any l. c. a. group, $\mu \in M(G)$, and $1 \leqslant p \leqslant \infty$, then μ defines an operator $C_\mu : L^p(G) \to L^p(G)$ by

$$\text{(i)} \qquad C_\mu f(g) = \mu * f(g) = \iint f(g_1^{-1}g)\,d\mu(g_1).$$

That C_μ is a bounded operator from $L^p(G)$ to $L^p(G)$ follows from the computation $(1/p + 1/q = 1)$:

$$\begin{aligned}\int \left|\int f(g_1^{-1}g)\,d\mu(g_1)\right|^p dm(g) &\leqslant \|\mu\|^{p/q}\iint |f(g_1^{-1}g)|^p\,d|\mu|(g_1)\,dm(g) \\ &= \|\mu\|^{p-1}\iint |f(g)|^p\,dm(g)\,d|\mu|(g_1) = \|\mu\|^p\,\|f\|_p^p;\end{aligned}$$

i. e. $\|\mu * f\|_p \leqslant \|\mu\|\,\|f\|_p$. Furthermore, since

$$(\mu * \nu) * f(g) = \int f(g_1^{-1}g)\, d\mu * \nu(g_1) = \iint f(g_2^{-1}g_1^{-1}g)\, d\mu(g_1)\, d\nu(g_2)$$
$$= \int \nu * f(g_1^{-1}g)\, d\mu(g) = \mu * (\nu * f),$$

we have $C_\mu C_\nu = C_{\mu * \nu}$ and $\mu \to C_\mu$ is a homomorphism of $M(G)$ into the algebra of bounded linear operators on $L^p(G)$. Clearly, $C_\delta = 1$ and so $\mu \to C_\mu$ maps the identity of $M(G)$ to the identity operator. Thus, the following is a trivial observation.

1.2.3. PROPOSITION. *If μ is an invertible element of $M(G)$, then C_μ is invertible on $L^p(G)$ for each p.*

The next proposition is not quite so trivial.

1.2.4. PROPOSITION. *Given any p with $1 \leqslant p \leqslant \infty$ and $\mu \in M(G)$, if C_μ is invertible on $L^p(G)$ then the Fourier transform $\hat{\mu}$ of μ is bounded away from zero on $\hat{G}$.*

PROOF. For $p = 2$ this follows from the fact that $f \to \hat{f} : L^2(G) \to L^2(\hat{G})$ is an isometry and $(\mu * f)\hat{} = \hat{\mu}\hat{f}$ (Rudin [5]). Thus, C_μ^{-1}, if it exists, is equivalent to the operator of multiplication by $\hat{\mu}^{-1}$ on $L^2(\hat{G})$. If this is to be a bounded operator, then $\hat{\mu}^{-1}$ must be bounded.

Note that for $1 \leqslant p \leqslant \infty$ and $1/p + 1/g = 1$, the adjoint map $C_\mu^* : L^q(G) \to L^q(G)$ of $C_\mu : L^p(G) \to L^p(G)$ is given by $T^{-1}C_\mu T$ acting on $L^q(G)$, where $Tf(g) = f(g^{-1})$. Since C_μ is invertible if and only if C_μ^* is invertible, it follows that C_μ is invertible on $L^p(G)$ if and only if it is invertible on $L^q(G)$.

The Riesz convexity theorem (Edwards [1]) implies that an operator defined on a space of functions dense in $L^p(G)$ for all p between p_1 and p_2, which is continuous in $L^{p_1}(G)$ norm and $L^{p_2}(G)$ norm, is also continuous in $L^p(G)$ norm for all p between p_1 and p_2. Applied to C_μ^{-1}, this forces C_μ to be invertible on $L^{p'}(G)$ for all p' between p and q, if it is invertible on $L^p(G)$. Since 2 is between p and q if $1/p + 1/q = 1$, we conclude that C_μ is invertible on $L^2(G)$ if it is invertible on $L^p(G)$ for any p. The proposition follows.

1.2.5. PROPOSITION. *For any p with $1 \leqslant p \leqslant \infty$ and for $\mu \in L(G) + \mathbf{C}\delta$, the operator C_μ on $L^p(G)$ is invertible if and only if $\hat{\mu}$ is nonvanishing on $\hat{G} \cup \{\infty\}$.*

PROOF. Since the spectrum of $L(G)$ is $\hat{G}$ and the Fourier transform $\mu \to \hat{\mu}$ is the Gelfand transform, the spectrum of $L(G) + \mathbf{C}\delta$ is the one point compactification, $\hat{G} \cup \{\infty\}$, of $\hat{G}$ and the Gelfand transform maps $\mu \in L(G) + \mathbf{C}\delta$ to $\hat{\mu}$ extended to $\hat{G} \cup \{\infty\}$ by continuity. Hence, $\mu \in L(G) + \mathbf{C}\delta$ is invertible if and only if $\hat{\mu}$ is nonvanishing on $\hat{G} \cup \{\infty\}$. Similarly, $\hat{\mu}$ is bounded away from zero on $\hat{G}$ if and only if it is nonvanishing on $\hat{G} \cup \{\infty\}$.

Thus, the conditions of 1.2.3 and 1.2.4 coincide for $\mu \in L(G) + \mathbf{C}\delta$ and yield a necessary and sufficient condition for C_μ to be invertible.

1.2.6. SINGULAR MEASURES. If $\mu \in M(G)$ but $\mu \notin L(G) + \mathbf{C}\delta$, the situation is considerably more complicated. For $p = 2$, the fact that C_μ is equivalent to multiplication by $\hat{\mu}$ on $L^2(\hat{G})$ means that C_μ is invertible if and only if $\hat{\mu}$ is bounded away from zero. As we shall see below, for $p = 1$ or ∞ the invertibility of C_μ is equivalent to the invertibility of μ in $M(G)$. Unfortunately, unlike the situation for $\mu \in L(G) + C\delta$, μ may fail to be invertible even though $\hat{\mu}$ is bounded away from zero (§1.3). Thus, for general $\mu \in M(G)$, the invertibility of C_μ may depend on p. Except in the case $p = 2$, there are no nice necessary and sufficient conditions that C_μ be invertible.

1.2.7. PROPOSITION (Wendel [1]). *The class of operators on $L^1(G)$ (or $C_0(G)$) of the form C_μ for $\mu \in M(G)$ is precisely the class of operators which commute with the translation operators T_g for $g \in G$ $(T_g f(g_1) = f(g^{-1}g_1))$.*

PROOF. Note that $T_g = C_{\delta_g}$, where $\delta_g \in M(G)$ is the point mass at g. Hence, each C_μ commutes with each T_g.

We prove the converse for C acting on either $L^1(G)$ or $C_0(G)$ by first considering any operator $C : C_0(G) \to C(G)$ commuting with translation. For such an operator we have $(Cf)(g) = (T_g^{-1}Cf)(0) = (CT_g^{-1}f)(0)$. Since $C_0(G)^* = M(G)$, there exists $\nu \in M(G)$ such that the continuous linear functional $f \to (Cf)(0)$ is determined by ν. Then

$$(Cf)(g) = (CT_g^{-1})f(0) = \int T_g^{-1} f(g_1)\, d\mu(g_1) = \int f(gg_1)\, d\nu(g_1).$$

Thus, $C = C_\mu$ if we set $\mu(E) = \nu(E^{-1})$.

Now if C is an operator on $L^1(G)$ which commutes with translations, then C^* on $L^\infty(G)$ commutes with translations also. Furthermore, C^* maps $C_0(G)$ into $C(G)$. In fact, for $f \in C_0(G)$, $g \to T_g^{-1} f$ is continuous from G into $L^\infty(G)$, and, hence, so is $g \to T_g^{-1}C^*f = C^*T_g^{-1}f$; it follows that C^*f is equal almost everywhere to an element of $C(G)$. We conclude from the previous paragraph, and the fact that $C_0(G)$ is weak-$*$ dense in $L^\infty(G)$, that $C^* = C_\nu$ for some $\nu \in M(G)$. It follows that $C = C_\mu$ for $\mu \in M(G)$ defined by $\mu(E) = \nu(E^{-1})$.

1.2.8. COROLLARY. *For $\mu \in M(G)$, the operator C_μ is invertible (on $L^1(G)$, $C_0(G)$, or $L^\infty(G)$) if and only if μ is invertible in $M(G)$.*

PROOF. Since C_μ commutes with translation, so must its inverse. On $L^1(G)$ or $C_0(G)$ we conclude from 1.2.7 that $C_\mu^{-1} = C_\nu$ for some $\nu \in M(G)$. Clearly ν must be an inverse in $M(G)$ for μ. Since C_μ on $L^\infty(G)$ has an inverse if and only if it has an inverse on $L^1(G)$, the proof is complete in all three cases.

1.2.9. REMARK. As we shall see, the problem of deciding when an arbitrary measure in $M(G)$ is invertible is quite difficult. In view of the identification of $M(G)$ with the algebra of translation invariant operators on $L^1(G)$ (or $C_0(G)$) expressed by 1.2.7, it is an embarrassment to be unable to decide whether or not a given measure in $M(G)$ is invertible.

In Chapter 9 we shall shed some light on this problem. However, a completely satisfactory solution has never been found (there may not be one).

§1.3. The spectrum of $M(G)$

1.3.1. If G is a nondiscrete l. c. a. group and Δ is the spectrum of $M(G)$, then $\hat{G}$ may be identified with a subset of Δ in such a way that if $\hat{\mu}$ is the Gelfand transform of μ, then the Fourier transform of μ is simply $\hat{\mu}$ restricted to $\hat{G}$. That is, we simply identify each $\gamma \in \hat{G}$ with the complex homomorphism $\mu \to \int \gamma \, d\mu$.

We cannot expect that Δ is just the one point compactification of $\hat{G}$ (as is the case for the spectrum of $L(G) + \mathbf{C}\delta$). In fact, the Fourier transform of a measure δ_g $(g \neq e)$ does not have a limit at infinity on $\hat{G}$. However, we might hope that $\hat{G}$ is at least dense in Δ. If this were true, then it would follow that $\mu \in M(G)$ would be invertible if and only if $\hat{\mu}$ were bounded away from zero on $\hat{G}$. The discovery that this is not true was one of the first indications of the complicated nature of the spectrum of $M(G)$. This and several other pathological facts concerning Δ are deduced from the following.

1.3.1. PROPOSITION (Williamson [1]). *If G is a nondiscrete l. c. a. group, then there exists a continuous, positive measure $\mu \in M(G)$ such that μ is symmetric $(\mu = \tilde{\mu})$ and its convolution powers, $\mu, \mu^2, \mu^3, \cdots$, are mutually singular (are supported on disjoint Borel sets).*

We shall not prove this here. There are several different constructions of such measures in the literature (cf. Weiner-Pitt [1], Sreider [2], Williamson [1], Hewitt [1], Hewitt-Kakutani [1]). Rudin [5] has a nice account of this and the other matters of this section.

1.3.2. PROPOSITION. *With μ as in (1.3.1) and $\|\mu\| = 1$, the spectrum of μ in $M(G)$ contains the entire unit circle, but the range of the Fourier transform of μ is real.*

PROOF. Since the measures $\mu, \mu^2, \mu^3, \cdots$ have disjoint supports, we have for any $z \in \mathbf{C}$ with $|z| = 1$:

$$\|(z\delta + \mu)^n\| = \left\| \sum_{p=0}^{n} \binom{n}{p} z^{n-p} \mu^p \right\| = \sum_{p=0}^{n} \binom{n}{p} \|z^{n-p}\mu^p\| = \sum_{p=0}^{n} \binom{n}{p} = 2^n.$$

It follows that $z\delta + \mu$ has spectral norm 2. Hence, there is an element $h \in \Delta$ such that $|z + \hat{\mu}(h)| = 2$. Note that h may be chosen from the Shilov boundary $\partial\Delta$, since the Gelfand transform of an element of $M(G)$ must assume its maximum modulus on $\partial\Delta$.

Since $\|\mu\| = 1$, we have $|\hat{\mu}(h)| \leqslant 1$. Since $|z| = 1$, we conclude that the equality $|z + \hat{\mu}(h)| = 2$ forces $\hat{\mu}(h) = z$. Thus, the spectrum of μ (range of $\hat{\mu}$) contains the entire unit circle.

On $\hat{G}$ we have $\hat{\tilde{\nu}} = \overline{\hat{\nu}}$ for any $\nu \in M(G)$. Since $\mu = \tilde{\mu}$, we conclude that $\hat{\mu}$ is real on $\hat{G}$.

We draw several conclusions from the above.

1.3.3. PROPOSITION. *If G is a nondiscrete l. c. a. group and Δ the spectrum of $M(G)$, then*

(a) *$\hat{G}$ is not dense in Δ; in fact, $\hat{G}$ is not even dense in the Shilov boundary $\partial\Delta$;*

(b) *$M(G)$ is not symmetric; that is, $\{\hat{\mu} \in C(\Delta); \mu \in M(G)\}$ is not closed under conjugation;*

(c) *there exist measures $\nu \in M(G)$ for which $\hat{\nu}$ is bounded away from zero on $\hat{G}$, but ν is not invertible in $M(G)$;*

(d) *given* $\epsilon > 0$, *there exists a measure* $\nu \in M(G)$ *such that* $|\hat{\nu}| < \epsilon$ *on* $\hat{G}$ *but* $\sup\{|\hat{\nu}(h)| : h \in \Delta\} = 1$.

PROOF. Part (a) follows immediately from 1.3.3. For part (b), note that if there were a measure $\nu \in M(G)$ with $\hat{\nu} = \overline{\hat{\mu}}$ on Δ, for the measure μ of 1.3.2, then $\hat{\nu}$ would agree with $\hat{\mu}$ on $\hat{G}$. Since the Fourier transform is injective on $M(G)$ (Rudin [5]), ν would have to be μ; since $\hat{\mu}$ assumes nonreal values on Δ, this is impossible.

For part (c), it suffices to let $\nu = i\delta + \mu$ for the measure μ of 1.3.2. For part (d) it suffices to choose ν to be an appropriately large power of $\omega = \frac{1}{2}(\delta - i\mu)$; in fact, $\hat{\omega}$ assumes the value 1 somewhere on Δ, but $|\hat{\omega}| \leqslant \sqrt{2}/2$ on $\hat{G}$ since $\hat{\mu}$ is real and $|\hat{\mu}| \leqslant 1$ on $\hat{G}$.

1.3.4. INDEPENDENT SETS. Some constructions of measures μ as in 1.3.1 yield measures with even more surprising properties.

An independent set in G is a set E such that $g_1, \cdots, g_k$ distinct points of E and $n_1, \cdots, n_k$ integers with $g_1^{n_1} g_2^{n_2} \cdots g_k^{n_k} = e$ imply that $g_1^{n_1} = g_2^{n_2} = \cdots = g_k^{n_k} = e$.

Sreider [2], in the case of the line, and Hewitt-Kakutani [1], in general, have shown that a nondiscrete l.c.a. group always contains an independent Cantor set. Hewitt and Kakutani [1] proved the following for such sets.

PROPOSITION. *If* $E \subset G$ *is an independent Cantor set, and* $M_c(E)$ *is the space of continuous measures supported on* E, *then each element* f *of* $M_c(E)^*$, *with* $\|f\| \leqslant 1$, *is the restriction to* $M_c(E)$ *of a complex homomorphism of* $M(G)$.

We shall not prove this here. A proof similar to Hewitt and Kakutani's appears in Rudin [5]. The key step is proving that if $\mu_1, \cdots, \mu_k \in M_c(E)$ are mutually singular, then so are all distinct products $\mu_1^{n_1} * \cdots * \mu_k^{n_k}$.

Rudin [4] shows that there is an independent Cantor set on the line such that $M_c(E)$ contains a positive measure with Fourier transform vanishing at infinity on $\hat{G}$. This shows that $M(G)$ can contain a measure μ with $\hat{\mu}$ vanishing at infinity on $\hat{G}$ but such that $\hat{\mu}$ does not vanish at all points of $\Delta\backslash\hat{G}$.

1.3.5. REMARKS. The preceding results indicate that the spectrum Δ of $M(G)$ is quite complicated, as is the problem of deciding when $\mu \in M(G)$ is invertible (i.e., when $\hat{\mu}$ does not vanish on Δ).

A description of Δ that is concrete enough to allow one to describe, for each given $\mu \in M(G)$, whether or not $\hat{\mu}$ vanishes on Δ, is probably not attainable. Sreider [1] gives a description of Δ in terms of what he calls generalized characters. However, this description is almost a restatement of the definition of complex homomorphism. The generalized characters of G are no easier to compute than are the complex homomorphisms of $M(G)$.

In Chapter 3 we shall prove that the spectrum of any measure algebra can be represented as the space $\hat{S}$ of semicharacters on some compact semigroup S. However, for $M(G)$ we have no concrete description of the corresponding S or $\hat{S}$. Hence, this also fails to give a really satisfactory description of the spectrum of $M(G)$. On the other hand, using this description of Δ as $\hat{S}$ and certain semigroup techniques, we are able in Chapter 9 to prove a theorem which significantly simplifies the problem of deciding when a measure in $M(G)$ is invertible. Basically, this theorem asserts that if μ^{-1} exists it must lie in a certain "small" subalgebra of $M(G)$. This reduces the invertibility problem in $M(G)$ to the same problem in an algebra which is far less complicated than $M(G)$.

§1.4. **Idempotents**

1.4.1. There is one well-known positive result regarding $M(G)$: this is the theorem of Cohen [1] which characterizes the idempotents of $M(G)$.

If A is a commutative Banach algebra with identity 1 and spectrum Δ, then for each idempotent $a = a^2 \in A$, the function $\hat{a}$ on Δ can assume only the values zero and one. Thus, $\hat{a}$ is the characteristic function of some open-compact subset of Δ. The converse is a deep theorem due to Shilov [1], which is proved using several complex variables and the several variable analytic functional calculus for Banach algebras.

1.4.2. THE GROUP $H^0(\Delta)$. We denote by $H^0(A)$ the additive subgroup of A generated by the idempotents of A. The additive group of continuous integer valued functions on Δ will be denoted $H^0(\Delta)$. Note that $a \to \hat{a}$ is a homomorphism of $H^0(A)$ into $H^0(\Delta)$. Actually,

PROPOSITION. *If A is a commutative Banach algebra with identity and if Δ is the spectrum of A, then $a \to \hat{a}$ is an isomorphism of $H^0(A)$ onto $H^0(\Delta)$.*

PROOF. An integer valued continuous function on a compact Hausdorff space must have the form $\Sigma_{i=1}^k n_i \chi_{E_i}$ for integers $n_1, \cdots, n_k$ and open-compact subsets $E_1, \cdots, E_k$. Hence it follows from Shilov's idempotent theorem (Shilov [1]) that a function of this form on Δ must be the Gelfand transform of an element $\Sigma_{i=1}^k n_i a_i \in A$ with $a_i^2 = a_i$. Hence, $a \to \hat{a}$ maps $H^0(A)$ onto $H^0(\Delta)$.

To complete the proof, we must prove that $a \to \hat{a}$ is injective on $H^0(A)$. If A is semisimple ($a \to \hat{a}$ is injective on A) there is no problem; however, we need not assume semisimplicity.

Suppose $a = n_1 a_1 + \cdots + n_k a_k$ with $a_i^2 = a_i \neq 0, n_i \neq 0$, and $a_i a_j = 0$ for $i \neq j$. Since $aa_i = n_i a_i$ we conclude that $\hat{a} = 0$ if and only if each $\hat{a}_i = 0$. However, since $a_i^n = a_i$ for each n, each a_i has spectral radius one. We conclude that $\hat{a} \neq 0$ under these conditions.

Now any element of $H^0(A)$ has the above form. In fact, if $a = m_1 b_1 + \cdots + m_l b_l$ with $b_i^2 = b_i$, then

$$1 = \prod_i [b_i + (1 - b_i)] = \sum_E c_E$$

where E is a subset of $\{1, \cdots, l\}$ and $c_E = \Pi_{j \in E} b_j \, \Pi_{j \notin E}(1 - b_j)$. Note that $c_E c_F = c_E$ if $E = F$ and $c_E c_F = 0$ if $E \neq F$. Also, for each i, $b_i c_E = 0$ or $b_i c_E = c_E$. It follows that a is a sum, with nonzero integral coefficients, of some subset of the set of elements c_E. It follows from the previous paragraph that $\hat{a} = 0$ implies $a = 0$.

1.4.3. THEOREM (Cohen [1]). *If G is an l. c. a. group, then the group $H^0(M(G))$ is generated by the idempotents in $M(G)$ which have the form $\gamma\mu$, where $\gamma \in \hat{G}$ and μ is Haar measure on some compact subgroup of G.*

We shall prove this in Chapter 8 using methods quite different from Cohen's. Our proof works equally well for other measure algebras.

Note that this theorem says that although we do not know exactly what Δ is for $M(G)$, we do know the extent to which it is disconnected. The degree of disconnectedness of Δ is measured by the group $H^0(\Delta)$ which is isomorphic to $H^0(M(G))$.

Note that, in general, 1.4.2 says that $H^0(A)$ depends only on the topology of Δ and not on A. In fact, $H^0(\Delta)$ is the zero dimensional Čech cohomology group of Δ with integral coefficients. We obtain our version of Cohen's idempotent theorem as a corollary of a general theorem which characterizes the Čech cohomology groups of all orders for the spectrum of any measure algebra.

§1.5. Wiener-Hopf equations

1.5.1. Let $R^+ = [0, \infty)$. For $k \in L^1(R)$, $\lambda \in \mathbf{C}$, the classical Wiener-Hopf equation for k and λ is

(i) $$\lambda f(x) + \int_0^\infty f(y)k(x-y)\,dy = h(x) \qquad (x \geqslant 0)$$

where $h \in L^p(R^+)$ and we seek a solution $f \in L^p(R^+)$. If we set $\mu = \lambda\delta + km$ (m Lebesgue measure), then this equation may be written as

(ii) $$W_\mu f(x) = \int_0^\infty f(y)\,d\mu(x-y) = h(x) \qquad (x \geqslant 0),$$

where the first equality defines the Wiener-Hopf operator $W_\mu : L^p(R^+) \to L^p(R^+)$.

Equation (ii) defines an operator W_μ for each $\mu \in M(R)$, but until recently the question of invertibility of W_μ could be handled only for $\mu \in L(R) + \mathbf{C}\delta$ or for μ a discrete measure. We shall discuss the general problem in Chapter 8 and obtain a complete solution in the case $p = 1$. Here we indicate the crux of the problem.

1.5.2. A DESCRIPTION OF W_μ. The operator W_μ can be described in terms of convolution operators in the following way. We consider $L^p(R^+)$ to be the subspace of $L^p(R)$ consisting of functions which vanish for $x < 0$. The operator W_μ acting on $f \in L^p(R^+)$ yields $PC_\mu f$, where $P: L^p(R) \to L^p(R^+)$ is the projection operator defined by $Pf(x) = f(x)$ for $x \geqslant 0$ and $Pf(x) = 0$ for $x < 0$.

The map $\mu \to C_\mu$ is a homomorphism of $M(R)$ into the algebra of bounded linear operators on $L^p(R)$. However, the map $\mu \to W_\mu$ is not a homomorphism. It is linear and bounded, but does not preserve multiplication. Thus, even if μ^{-1} exists in $M(R)$, $W_{\mu^{-1}}$ may not be an inverse for W_μ. However, one can determine sufficient conditions that W_μ be invertible by noting that $\mu \to W_\mu$ does preserve certain kinds of products.

We let $M(R^+)$ and $M(R^-)$ denote the subalgebras of $M(R)$ consisting of measures concentrated on $R^+ = [0, \infty)$ and $R^- = (-\infty]$, respectively.

1.5.3. LEMMA. *For* $\mu, \nu \in M(R)$ *we have* $W_{\mu * \nu} = W_\mu W_\nu$ *in each of the following two cases*:

(a) $\nu \in M(R^+)$, μ *arbitrary, and*

(b) $\mu \in M(R^-)$, ν *arbitrary.*

PROOF. If $\nu \in M(R^+)$ then $C_\nu f = \nu * f \in L^p(R^+)$ for each $f \in L^p(R^+)$. Hence, $W_\nu f = PC_\nu f = C_\nu f$ and $W_\mu W_\nu f = PC_\mu C_\nu f = PC_{\mu * \nu} f = W_{\mu * \nu} f$.

Now if $\mu \in M(R^-)$, then for $f \in L^p(R^+)$ we have $W_\mu W_\nu f = PC_\mu PC_\nu f = PC_\mu [C_\nu f + (P-1)C_\nu f]$. However, $(P-1)C_\nu f \in L^p(R^-)$ and so $C_\mu (P-1)C_\nu f \in L^p(R^-)$ as well. We conclude that $PC_\mu (P-1)C_\nu f = 0$ and $W_\mu W_\nu f = PC_\mu C_\nu f = PC_{\mu*\nu} f = W_{\mu*\nu} f$.

1.5.4. FACTORIZATION AND LOGARITHMS. Suppose $\mu \in M(R)$ has a factorization of the form $\mu = \mu_- * \mu_+$ with μ_- an invertible element of the algebra $M(R^-)$ and μ_+ an invertible element of the algebra $M(R^+)$. Then by 1.5.3, we have $W_\mu = W_{\mu_-} W_{\mu_+}$, $W_{\mu_-} W_{\mu_-^{-1}} = W_{\mu_-^{-1}} W_{\mu_-} = W_{\mu_-^{-1} * \mu_-} = W_\delta = 1$, and $W_{\mu_+} W_{\mu_+^{-1}} = W_{\mu_+^{-1} * \mu_+} = W_\delta = 1$. Since W_μ is the product of two invertible operators, it is also invertible.

One way to ensure that μ has a factorization of the above form is the following: Suppose $\mu = e^\nu$ for some $\nu \in M(R)$. If we set $\nu_+ = \nu|_{R^+}$ and $\nu_- = \nu - \nu_+$, then $\mu_+ = e^{\nu_+}$ and $\mu_- = e^{\nu_-}$ are invertible elements of $M(R^+)$ and $M(R^-)$, respectively, and $\mu = \mu_- * \mu_+$. Hence,

1.5.5. PROPOSITION. *If* $\mu \in M(R)$ *has a logarithm in* $M(R)$ ($\mu = e^\nu$ *for some* $\nu \in M(R)$), *then the Wiener-Hopf operator* W_μ *is invertible on* $L^p(R^+)$ *for all* p *with* $1 \leqslant p \leqslant \infty$.

It turns out that for $p = 1$ the converse is also true (Chapter 8).

In order to apply this proposition, we need methods for determining when a measure in $M(R)$ has a logarithm.

§1.6. Logarithms

1.6.1. If A is a commutative Banach algebra with identity, we denote by A^{-1} the multiplicative group of all invertible elements of A. If $a = e^b$ for some $b \in A$, then $a \in A^{-1}$ with $a^{-1} = e^{-b}$. Hence, the group $\exp(A)$ of all elements of A with logarithms in A is a subgroup of A^{-1}.

It is not difficult to see that if A^{-1} is given the norm topology, then it is a topological group with $\exp(A)$ as an open subgroup. That $\exp(A)$ is open follows from the fact that if $\|1 - a\| < 1$, then the convergent power series $-\sum_{n=1}^\infty n^{-1}(1-a)^n$ yields a logarithm for a.

Note that if $a = e^b \in \exp(A)$ then $t \to e^{tb}$ ($t \in [0, 1]$) yields an arc in $\exp(A)$ connecting a to 1. Hence, $\exp(A)$ is exactly the connected component of the identity in A^{-1}. We denote the discrete group $A^{-1}/\exp(A)$ by $H^1(A)$.

1.6.2. THE ARENS-ROYDEN THEOREM. *Let* Δ *be the spectrum of* A. *We denote the group* $H^1(C(\Delta)) = C(\Delta)^{-1}/\exp(C(\Delta))$ *by* $H^1(\Delta)$.

Note that the Gelfand transform $a \to \hat{a} : A \to C(\Delta)$ maps A^{-1} into $C(\Delta)^{-1}$ and $\exp(A)$ into $\exp(C(\Delta))$. In fact, for any identity preserving Banach algebra homomorphism $\phi : A \to B$ it is obvious that $\phi A^{-1} \subset B^{-1}$ and $\phi \exp(A) \subset \exp(B)$; thus ϕ induces a map of $H^1(A)$ into $H^1(B)$. In particular, $a \to \hat{a}$ induces a homomorphism of $H^1(A)$ into $H^1(\Delta)$.

THEOREM (ARENS [1] AND ROYDEN [1]). *If A is a commutative Banach algebra with identity 1 and spectrum Δ, then the map $H^1(A) \to H^1(\Delta) = H^1(C(\Delta))$, induced by the Gelfand transform, is an isomorphism.*

We shall not prove this theorem here. The proof is quite complicated and depends on the theory of several complex variables and the several variable analytic functional calculus. A very nice proof appears in Gamelin [1].

1.6.3. COHOMOLOGY. We have now seen two groups $H^0(A)$ and $H^1(A)$, which are important in the study of a Banach algebra A. In both cases, these groups are isomorphic to groups $H^0(\Delta)$ and $H^1(\Delta)$, which can be defined purely in terms of the topology of the spectrum Δ. In fact, these groups can be defined for any topological space Δ.

If Δ is a Hausdorff space, we define $H^0(\Delta)$ to be the additive subgroup of $C(\Delta)$ consisting of integer valued continuous functions, and $H^1(\Delta)$ to be $C(\Delta)^{-1}/\exp C(\Delta)$. Note that these groups are related in the following way: they are the kernel and cokernel, respectively, of the map $f \to \exp(2\pi i f) : C(\Delta) \to C(\Delta)^{-1}$.

It turns out that if one uses either sheaf or Čech cohomology, then $H^0(\Delta)$ and $H^1(\Delta)$ are the zero and one-dimensional cohomology groups of Δ with integral coefficients.

In Chapters 6 and 7 we shall use techniques of cohomology theory to compute these groups in the case that Δ is the spectrum of a measure algebra. This leads to a proof of Cohen's idempotent theorem for general measure algebras and to a characterization of those invertible elements of a measure algebra which have logarithms (Chapter 8).

1.6.4. THE CASE OF $L(R) + \mathbf{C}\delta$. In order to illustrate the connection between $H^1(A)$ and the cohomology of Δ, we compute $H^1(A)$ for the algebra $L(R) + \mathbf{C}\delta$. Note that we expect to get a group isomorphic to the integers since Δ is the one point compactification, $R \cup \{\infty\}$, of R in this case.

Now if f is an element of $C(R \cup \{\infty\})^{-1}$ then f is nonvanishing on $R \cup \{\infty\}$, and a logarithm for f must, up to a constant, have the form

(i) $$g(x) = \int_0^x f'(y) f^{-1}(y)\, dy.$$

The function g defined by (i) will always be continuous and bounded on R. However, it may fail to have a limit at ∞. In general, the limits $\lim_{x\to\infty} g(x)$ and $\lim_{x\to-\infty} g(x)$ exist and

$$\omega(f) = \frac{1}{2\pi i}\left[\lim_{x\to\infty} g(x) - \lim_{x\to-\infty} g(x)\right]$$

is an integer (the winding number of f about zero). Obviously, f has a logarithm in $C(R \cup \{\infty\})$ if and only if $\omega(f) = 0$.

It is not difficult to see that $\omega(f_1f_2) = \omega(f_1) + \omega(f_2)$ for $f_1, f_2 \in C(R \cup \infty)$ and that the function $x \to (1 + ix)(1 - ix)^{-1}$ has winding number one. It follows that every integer is the winding number of some function in $C(R \cup \{\infty\})^{-1}$ and that $\omega : C(R \cup \{\infty\})^{-1} \to Z$ is a surjective homomorphism with kernel $\exp(C(R \cup \{\infty\}))$. Thus, ω induces an isomorphism of $H^1(R \cup \{\infty\})$ onto Z. In view of the Arens-Royden Theorem 1.6.3 and the fact that $R \cup \{\infty\}$ is the spectrum, and $\mu \to \hat{\mu}$ the Gelfand transform for $L(R) + \mathbf{C}\delta$, we have

PROPOSITION. *The group* $H^1(L(R) + \mathbf{C}\delta)$ *is isomorphic to the group of integers. In fact, the map* $\mu \to \omega(\hat{\mu})$ *is a homomorphism of* $(L(R) + \mathbf{C}\delta)^{-1}$ *onto* Z *with kernel* $\exp(L(R) + \mathbf{C}\delta)$.

Note that, in view of 1.5.5, we have

COROLLARY. *If* $\mu \in L(R) + \mathbf{C}\delta$ *then the Wiener-Hopf operator* W_μ *has an inverse if* $\hat{\mu}$ *is nonvanishing on* $R \cup \{\infty\}$ *and has winding number zero about the origin.*

§1.7. **Other measure algebras**

1.7.1. So far, our discussion has been limited to the algebras $L(G)$ and $M(G)$. However, other kinds of measure algebras occur in many interesting contexts.

Measure algebras on finite semigroups were studied by Hewitt and Zuckermann [1]. Hewitt and Williamson [1] studied the measure algebra on the multiplicative nonnegative integers. A particularly interesting class of measure algebras was introduced by Arens and Singer [1]. A typical algebra of this type is described as $L(T)$, the subalgebra of $L(G)$ consisting of measures concentrated on a closed subsemigroup $T \subset G$ with dense interior. Measure algebras on idempotent semigroups were studied by Newman [1], [2]. Ross [1] studied the measure algebra of a totally ordered set under max multiplication. We shall discuss each of these examples in more detail in Chapter 4.

The problems of determining the spectrum Δ and characterizing the idempotents, invertible elements, and elements with logarithms, are important for measure algebras other than $M(G)$. Furthermore, the techniques we shall use to attack these problems work as well for general measure algebras as they do for $M(G)$.

1.7.2. SEMICHARACTERS. We noted in §1.1 that if S is a locally compact semigroup, then the semicharacters in $\hat{S}$ determine complex homomorphisms of $M(S)$ and, hence, of any subalgebra of $M(S)$. In the case of $L(G)$, the points of $\hat{G}$ determine all the complex homomorphisms. Similarly, $\hat{T}$ determines the entire spectrum for measure algebras on a discrete semigroup T, and for the algebras $L(T)$ of Arens and Singer (Chapter 4). This is not true of all measure algebras ($M(G)$ for nondiscrete G is a counterexample). However, we shall show in Chapter 3 that every measure algebra $\mathfrak{M}$ can be embedded in an algebra $M(S)$ for which $\hat{S}$ is the spectrum of $\mathfrak{M}$. This allows us to use the special properties of a space of semicharacters in attacking the problems we have been discussing.

1.7.3. ANALYTIC STRUCTURE, BOUNDARIES. A simple algebra which exhibits much of the behavior we shall come to expect in general measure algebras is the algebra

$M(Z^+) = l^1(Z^+)$ on the additive semigroup Z^+ of nonnegative integers. This is both a discrete semigroup algebra and an Arens-Singer algebra. Its spectrum is the space of bounded semicharacters on Z^+. We can identify this space with the unit disc $D \subset \mathbf{C}$, where $z \in D$ determines the semicharacter $n \to z^n$.

The Gelfand transform for $l^1(\mathbf{Z}^+)$ is the map $\mu \to \hat{\mu}$, where

$$\hat{\mu}(z) = \sum_{n=0}^{\infty} \mu(\{n\}) z^n .$$

Thus, the image of the Gelfand transform for this algebra consists of those analytic functions on D with absolutely convergent power series.

Note that the Shilov boundary for this algebra consists of the unit circle and corresponds to the set of semicharacters on Z^+ with modulus one. These points of the spectrum are also the one-point Gleason parts. The remainder of the spectrum is a single Gleason part and has analytic structure.

For general measure algebras, analytic structure in the spectrum $\hat{S}$ arises from the existence of semicharacters $f \in \hat{S}$ for which $|f|$ has values other than zero or one. This is due to the fact that $|f|^z$ is also a semicharacter in $\hat{S}$ for $z \in \mathbf{C}$ with $\operatorname{Re} z > 0$, and the fact that each $f \in \hat{S}$ has a factorization of the form $f = h|f|$ (Chapter 3).

In Chapter 10 we shall study analytic structure, Gleason parts, and Shilov and Strong boundaries for measure algebras. The semicharacter representation of the spectrum turns out to have great value in this endeavor.

CHAPTER 2. *L*-SPACES

Measure algebras are special Banach spaces of measures with a special kind of multiplication. In this chapter, we characterize these special spaces of measures and study the properties which make them special.

The class we are interested in is the class of L-spaces introduced by Kakutani [1]. Kakutani defines an L-space to be a certain kind of Banach vector lattice. However, in order to minimize the amount of background machinery we rely on, we shall give a concrete definition and development which relies only on standard measure theory.

§2.1. *L*-Subspaces of a measure space

2.1.1. Let (X, Σ) be a measurable space, that is, a set X together with a σ-algebra Σ of subsets of X. We denote by $M(X, \Sigma)$ the Banach space of all complex valued, finite, countably additive measures on Σ. For $\mu \in M(X, \Sigma)$ we denote the total variation measure for μ by $|\mu|$ and the total variation norm, $|\mu|(X)$, by $\|\mu\|$.

We shall use the notation "$\nu << \mu$" to denote the statement that $\nu \in M(X, \Sigma)$ is absolutely continuous with respect to $\mu \in M(X, \Sigma)$. Similarly, "$\mu \perp \nu$" will mean that μ and ν are mutually singular. The Radon-Nikodym theorem states that $\nu << \mu$ if and only if $\nu = f\mu$ for some $f \in L^1(\mu)$, where $f\mu(E) = \int \chi_E f \, d\mu$.

We denote the space $\{\nu \in M(X, \Sigma) : \nu << \mu\}$ by $L(\mu)$. Note that it is isometric as a Banach space to $L^1(\mu)$.

2.1.2. L-SUBSPACES. A closed linear subspace $\mathfrak{M} \subset M(X, \Sigma)$ will be called an L-subspace if $\mu \in \mathfrak{M}$, $\nu \in M(X, \Sigma)$, and $\nu << \mu$ imply that $\nu \in \mathfrak{M}$.

Note that for $\mu \in M(X, \Sigma)$, $\mathrm{Re}(\mu)$, $\mathrm{Im}(\mu)$, and $|\mu|$ are all absolutely continuous with respect to μ. Hence, if $\mathfrak{M}$ is an L-subspace, then it contains $\mathrm{Re}(\mu)$, $\mathrm{Im}(\mu)$, and $|\mu|$ for each $\mu \in \mathfrak{M}$.

2.1.3. THE ORDER RELATION. If $\mathfrak{M} \subset M(X, \Sigma)$ is an L-subspace and $\mathfrak{M}_r$ is the real subspace of $\mathfrak{M}$ consisting of real valued measures, then $\mathfrak{M}_r$ is a real ordered Banach space under the order relation: $\mu \leqslant \nu$ if and only if $\mu(E) \leqslant \nu(E)$ for all $E \in \Sigma$. In fact, $\mathfrak{M}_r$ is a vector lattice in the sense that for each pair $\mu, \nu \in \mathfrak{M}_r$, there is a unique least upper bound $\mu \vee \nu$ and greatest lower bound $\mu \wedge \nu$. To see this, we simply set $\omega = |\mu| + |\nu|$, $\mu = f\omega$, and $\nu = g\omega$ for real functions $f, g \in L^1(\omega)$. Then $\mu \vee \nu = (f \vee g)\omega$ and $\mu \wedge \nu = (f \wedge g)\omega$.

If $\{\mu_\alpha\}$ is a family of measures in $\mathfrak{M}_r$, each of which is bounded by a fixed measure $\mu \in \mathfrak{M}_r$, then there is a least upper bound $\vee\mu_\alpha \in \mathfrak{M}_r$. In fact, for a countable family

$\{\mu_\alpha\}, \bigvee \mu_\alpha$ can be obtained as $f\mu$, where $f = \bigvee_\alpha f_\alpha$ and $\mu_\alpha = f_\alpha \mu$ for each α (note $f_\alpha \leqslant 1$). Using the fact that μ is finite, it is not difficult to see that even if $\{\mu_\alpha\}$ is not countable it contains a countable subset whose supremum dominates each μ_α. Hence, arbitrary bounded sets have least upper bounds. A vector lattice with this property is called complete.

2.1.4. INVARIANCE. Certain measure theoretic operations and relations in an L-subspace $\mathfrak{M} \subset M(X, \Sigma)$ are determined entirely by the linear structure, order relation and norm in $\mathfrak{M}$. These are the operations and relations that will be preserved by any order-preserving isometry between two such spaces. We shall describe such things as being invariant.

PROPOSITION. *The relations* $\mu \perp \nu$ *and* $\nu << \mu$, *and the operations* $\mu \to |\mu|$ *and* $\mu \to \bar{\mu}$ *are invariant.*

PROOF. Since $\mathfrak{M}_r = \mathfrak{M}^+ - \mathfrak{M}^+$ $(\mathfrak{M}^+ = \{\mu \in \mathfrak{M} : \mu \geqslant 0\})$ we have that $\mathfrak{M}_r$ and, hence, $\mu \to \bar{\mu}$ are invariant. The relation $\mu \perp \nu$ holds if and only if $|\mu| \wedge |\nu| = 0$, while $\nu << \mu$ holds if and only if $|\nu| = \lim_{t \to \infty} (|\nu| \wedge t|\mu|)$. Hence, these relations will be invariant if $\mu \to |\mu|$ is. However, one easily checks that $|\mu| = \bigvee \{\mathrm{Re}(\lambda\mu) : |\lambda| = 1, \lambda \in \mathbf{C}\}$. Thus, $\mu \to |\mu|$ is also invariant.

§2.2. *L*-homomorphisms

2.2.1. DEFINITION. Let $\mathfrak{M}_1 \subset M(X_1, \Sigma_1)$ and $\mathfrak{M}_2 \subset M(X_2, \Sigma_2)$ be L-subspaces and $\phi : \mathfrak{M}_1 \to \mathfrak{M}_2$ a bounded linear map. We shall call ϕ an L-homomorphism if $0 \leqslant \mu \in \mathfrak{M}$ implies:

(a) $\phi\mu \geqslant 0$;

(b) $\|\phi\mu\| = \|\mu\|$;

(c) $\phi\{\nu \in \mathfrak{M}_1 : 0 \leqslant \nu \leqslant \mu\} = \{\omega \in \mathfrak{M}_2 : 0 \leqslant \omega \leqslant \phi\mu\}$.

In a certain sense, it turns out to be true that L-homomorphisms are the maps of measure spaces which are induced by underlying point maps.

2.2.2. PROPOSITION. *For measure spaces* $M(X_1, \Sigma_1)$ *and* $M(X_2, \Sigma_2)$, *and* $\alpha : X_1 \to X_2$ *any measurable transformation* $(\alpha^{-1}(E) \in \Sigma_1$ *for each* $E \in \Sigma_2)$, *the map* $\mu \to \mu \circ \alpha^{-1} : M(X_1, \Sigma_1) \to M(X_2, \Sigma_2)$ *is an* L-*homomorphism, as is its restriction to any* L-*subspace of* $M(X_1, \Sigma_1)$.

PROOF. If $\mu \in M(X_1, \Sigma_1)$ and $\mu \geqslant 0$, then obviously $\mu \circ \alpha^{-1} \geqslant 0$ and $\|\mu \circ \alpha^{-1}\| = \mu \circ \alpha^{-1}(X_2) = \mu(X_1) = \|\mu\|$. Thus, (a) and (b) of 2.2.1 hold.

To verify (c), we must show that if $\mu \geqslant 0$ and $0 \leqslant \omega \leqslant \mu \circ \alpha^{-1}$, then $\omega = \nu \circ \alpha^{-1}$ for some $\nu \in M(X_1, \Sigma_1)$ with $0 \leqslant \nu \leqslant \mu$. However, it suffices to choose $\nu = (f \circ \alpha)\mu$, where $\omega = f(\mu \circ \alpha^{-1})$. In fact, $0 \leqslant f \leqslant 1$, and

$$\begin{aligned}((f \circ \alpha)\mu) \circ \alpha^{-1}(E) &= \int \chi_{\alpha^{-1}(E)}(x) f(\alpha(x))\, d\mu(x) \\ &= \int \chi_E(\alpha(x)) f(\alpha(x))\, d\mu(x) = \int \chi_E(y) f(y)\, d\mu \circ \alpha^{-1}(y) \\ &= f(\mu \circ \alpha^{-1})(E) = \omega(E).\end{aligned}$$

Hence, $\mu \to \mu \circ \alpha^{-1}$ is an L-homomorphism. Clearly, the restriction of an L-homomorphism to an L-subspace is also an L-homomorphism.

2.2.3. LEMMA. *If* $\phi : \mathfrak{M}_1 \to \mathfrak{M}_2$ *is an L-homomorphism,* $0 \leqslant \mu \in \mathfrak{M}_1$, *and* $\phi\mu = \omega_1 + \omega_2$ *with* $\omega_1 \perp \omega_2$, *then there exist unique positive measures* $\nu_1 \leqslant \mu, \nu_2 \leqslant \mu$, *with* $\omega_i = \phi\nu_i$. *Necessarily,* $\mu = \nu_1 + \nu_2$ *and* $\nu_1 \perp \nu_2$.

PROOF. Note that 2.2.1(c) implies the existence of measures ν_1, ν_2 with $0 \leqslant \nu_i \leqslant \mu$ and $\phi\nu_i = \omega_i$. Since $\omega_1 \perp \omega_2$ and $\omega_i \geqslant 0$, we have $\omega_1 \wedge \omega_2 = 0$. This implies that $\nu_1 \wedge \nu_2 = 0$; otherwise, $\phi(\nu_1 \wedge \nu_2)$ would be nonzero (by 2.2.1(b)) and dominated by both ω_1 and ω_2 (by 2.2.1(a)). Since $\nu_1 \wedge \nu_2 = 0$ and $\nu_1 \leqslant \mu, \nu_2 \leqslant \mu$, we have $\nu_1 + \nu_2 \leqslant \mu$. However, the fact that $\phi(\mu - \nu_1 - \nu_2) = \phi\mu - \omega_1 - \omega_2 = 0$ implies that $\nu_1 + \nu_2 = \mu$.

The uniqueness of ν_1 and ν_2 proceeds as follows: if we fix a choice for ν_1, then for any possible choice for ν_2 we must have $\nu_1 + \nu_2 = \mu$. Hence, there is only one possible choice, $\mu - \nu_1$, for ν_2. Similarly, ν_1 is unique.

2.2.4. PROPOSITION. *Let* $\phi : \mathfrak{M}_1 \to \mathfrak{M}_2$ *be an L-homomorphism and* $0 \leqslant \mu \in \mathfrak{M}_1$. *Then* ϕ *maps* $L(\mu)$ *into* $L(\phi\mu)$.

PROOF. For a measure in $L(\mu)$ of the form $\nu = \Sigma \lambda_i \mu_i$ $(\lambda_i \in \mathbf{C}, 0 \leqslant \mu_i \leqslant \mu)$, we have $0 \leqslant \phi\mu_i \leqslant \phi\mu$ by 2.2.1(a) and, hence, $\phi\nu \in L(\phi\mu)$. Since measures of this form are dense in $L(\mu)$ we conclude that ϕ maps $L(\mu)$ into $L(\phi\mu)$.

2.2.5. PROPOSITION. *If* μ *and* ω *are positive measures and* $\phi : L(\mu) \to L(\omega)$ *is an L-homomorphism with* $\phi\mu = \omega$, *then the adjoint map* $\phi^* : L^\infty(\omega) \to L^\infty(\mu)$ *is a Banach algebra homomorphism which maps the identity to the identity.*

PROOF. The identity 1 of an algebra $L^\infty(\mu)$ is characterized, as a linear functional on $L(\mu)$, by the condition $(1, \nu) = \int 1\, d\nu = \|\nu\|$ for $0 \leqslant \nu \in L(\mu)$. Thus, it follows from 2.2.1(b) that $(\phi^*1, \nu) = (1, \phi\nu) = \|\phi\nu\| = \|\nu\| = (1, \nu)$ for $0 \leqslant \nu \in L(\mu)$ and, hence, that $\phi^*1 = 1$.

In order to prove that ϕ^* preserves pointwise multiplication of L^∞ functions, we first use ϕ and Lemma 2.2.3 to construct a map $E \to E'$ from ω-measurable sets to μ-measurable sets.

If E is any ω-measurable set and $\widetilde{E}$ is its complement, we set $\omega_1 = \chi_E\omega = \omega|_E$ and $\omega_2 = \chi_{\widetilde{E}}\,\omega = \omega|_{\widetilde{E}}$. Since $\omega_1 \perp \omega_2$ and $\omega_1 + \omega_2 = \omega$, Lemma 2.2.3 implies that there are a unique ν_1 and a unique ν_2 with $0 \leqslant \nu_i \leqslant \mu$ and $\phi\nu_i = \omega_i$. Furthermore, $\nu_1 \perp \nu_2$ and $\nu_1 + \nu_2 = \mu$. We conclude that there is a unique (up to sets of μ-measure zero) μ-measurable set E' with $\nu_1 = \chi_{E'}\mu$ and $\nu_2 = \chi_{(E')^\sim}\mu$, where $(E')^\sim$ is the complement of E'. This defines a correspondence $E \to E'$ between ω-measurable sets and μ-measurable sets. Note that, by the symmetry of the construction, $\widetilde{E}' = (E')^\sim$.

We claim that a measure ρ, with $0 \leqslant \rho \leqslant \mu$, is concentrated on E' if and only if $\phi\rho$ is concentrated on E. Clearly, if ρ is concentrated on E', then $\rho \leqslant \nu_1 = \chi_{E'}\mu$ and

$\phi\rho \leqslant \omega_1$ is concentrated on E. Conversely, if $\phi\rho$ is concentrated on E, then $\phi\rho \perp \omega_2$, $\rho \leqslant \rho + \nu_2$ and $\nu_2 \leqslant \rho + \nu_2$; thus, Lemma 2.2.3 with μ replaced by $\rho + \nu_2$ implies that $\rho \perp \nu_2$, i.e., that ρ is concentrated on E'. This establishes the claim.

We now prove that $(E \cap F)' = E' \cap F'$. By the above, $(E \cap F)'$ is characterized up to sets of measure zero by the fact that for $0 \leqslant \rho \leqslant \mu$, ρ is concentrated on $(E \cap F)'$ if and only if $\phi\rho$ is concentrated on $E \cap F$. However, it is apparent that, for $0 \leqslant \rho \leqslant \mu$, ρ is concentrated on $E' \cap F'$ if and only if $\phi\rho$ is concentrated on E and on F, i.e., on $E \cap F$. Hence, $(E \cap F)' = E' \cap F'$.

The proof that ϕ^* is a homomorphism now proceeds as follows: the sets E' and E are related by $\phi^*\chi_E = \chi_{E'}$; this follows from the fact that, for ρ concentrated on E,

$$\begin{aligned}(\phi^*\chi_E, \rho) = (\chi_E, \phi\rho) &= \int \chi_E \, d\phi\rho = \int 1 \, d\phi\rho \\ &= \int \phi^* 1 \, d\rho = \int 1 \, d\rho = \int \chi_{E'} \, d\rho = (\chi_{E'}, \rho);\end{aligned}$$

while for ρ concentrated on $(E')^\sim$,

$$\begin{aligned}(\phi^*\chi_E, \rho) = (\chi_E, \phi\rho) &= \int \chi_E \, d\phi\rho = 0 \\ &= \int \chi_{E'} \, d\rho = (\chi_{E'}, \rho).\end{aligned}$$

Since $E \to E'$ preserves intersections, ϕ^* preserves products of simple functions. Since the simple functions are dense in $L^\infty(\omega)$, ϕ^* is a homomorphism.

§2.3. *L*-spaces and duals of *L*-spaces

2.3.1. Kakutani [1] defines an abstract (real) L-space $\mathfrak{M}$ to be a real Banach vector lattice in which the conditions $\|\mu + \nu\| = \|\mu\| + \|\nu\|$ for $\mu, \nu \geqslant 0$ and $\|\mu - \nu\| = \|\mu\| + \|\nu\|$ for $\mu \wedge \nu = 0$ are met. An abstract complex L-space $\mathfrak{M}$ is defined to be a complex Banach space with a distinguished real subspace $\mathfrak{M}_r$ such that $\mathfrak{M} = \mathfrak{M}_r \oplus i\mathfrak{M}_r$, $\mathfrak{M}_r$ is a real L-space, and for each $\mu \in \mathfrak{M}$ the element $\bigvee\{\mathrm{Re}(\lambda\mu) : |\lambda| = 1, \lambda \in \mathbf{C}\} = |\mu|$ has the same norm as μ (Rieffel [1]).

Kakutani [1] proves that an abstract real L-space has a norm and order preserving representation as an L-subspace of a real measure space $M_r(X, \Sigma)$. Rieffel [1] proves the analogous result for complex L-spaces.

There is a class of vector lattices called M-spaces (Kakutani [2]) which is dual to the class of L-spaces. The dual of an L-space is an M-space and the dual of an M-space is an L-space. Every M-space may be represented as the space $C(X)$ for an appropriate compact Hausdorff space X.

A nice discussion of these matters appears in Schaeffer [1]. However, our approach will be somewhat different. In order to avoid the machinery of vector lattices, we give a definition of L-spaces which is more concrete and relies, instead, on the machinery of measure theory.

2.3.2. Definition. An ordered, complex Banach space $\mathfrak{M}$ will be called an L-space if it is isomorphic and isometric, in an order preserving fashion, to an L-subspace of some measure space $M(X, \Sigma)$.

Recall that the relations "$\mu << \nu$", "$\mu \perp \nu$", and operations $\mu \to \bar{\mu}, \mu \to |\mu|$ can be defined in an L-subspace of $M(X, \Sigma)$ using only the order relation, norm, and linear structure (2.1.4). Hence, these things are intrinsic to an L-space $\mathfrak{M}$ and do not depend on the way in which $\mathfrak{M}$ is represented as an L-subspace of a measure space.

In particular, for $\mu \in \mathfrak{M}$, the subspace $L(\mu)$ is defined independently of the representation. Now the dual, $L(\mu)^*$, of $L(\mu)$ has a representation as the Banach algebra $L^\infty(\mu_X)$ for any concrete representation $\mu \to \mu_X : \mathfrak{M} \to M(X, \Sigma)$ of $\mathfrak{M}$. It is conceivable that the algebraic structure of this algebra could depend on the representation. However

2.3.3. PROPOSITION. *If* $\mathfrak{M}$ *is an* L-*space and* $\mu \to \mu_X : \mathfrak{M} \to M(X, \Sigma)$ *and* $\mu \to \mu_Y : \mathfrak{M} \to M(Y, \Sigma')$ *are concrete representations of* $\mathfrak{M}$, *then the adjoint of* $\mu_X \to \mu_Y : L(\nu_X) \to L(\nu_Y)$ *is an isomorphism between the algebras* $L^\infty(\nu_Y)$ *and* $L^\infty(\nu_X)$ *for each* $\nu \in \mathfrak{M}$.

PROOF. A norm and order preserving isomorphism between two L-subspaces is clearly an L-homomorphism in each direction. Hence, it follows from 2.2.5 that the dual of $\mu_X \to \mu_Y : L(\nu_X) \to L(\nu_Y)$ is an algebraic isomorphism.

Henceforth, for an L-space $\mathfrak{M}$ and $\mu \in \mathfrak{M}, L^\infty(\mu)$ will denote the adjoint of $L(\mu)$ with the unique Banach algebra structure it inherits from any concrete representation of $\mathfrak{M}$.

Note that there is a natural involution $f \to \bar{f}$ on $L^\infty(\mu)$. This can be defined abstractly by $(\bar{f}, \nu) = \overline{(f, \bar{\nu})}$. In any concrete representation this operation coincides with complex conjugation.

The algebra $L^\infty(\mu)$ is a commutative C^*-algebra with identity; that is, it satisfies $\|\bar{f}f\| = \|f\|^2$.

2.3.4. THEOREM. *If* $\mathfrak{M}$ *is an* L-*space, there is a unique commutative* C^*-*algebra structure on* $\mathfrak{M}^*$ *such that each of the restriction maps* $f \to f_\mu : \mathfrak{M}^* \to L^\infty(\mu)$, *for* $\mu \in \mathfrak{M}$, *is an involution preserving homomorphism. Furthermore,* $\mathfrak{M}^*$ *has an identity* 1 *and* $1_\mu = 1 \in L^\infty(\mu)$ *for each* $\mu \in \mathfrak{M}$.

PROOF. For each $f \in \mathfrak{M}^*$ and $\mu \in \mathfrak{M}$, let $f_\mu \in L^\infty(\mu)$ be the restriction of f to $L(\mu)$. Thus, if we consider $\mathfrak{M}$ to be an L-subspace of $M(X, \Sigma)$ via any concrete representation, then $(f, \nu) = \int f_\mu \, d\nu$ for all $\nu << \mu$. Note that $\omega << \mu$ implies that $f_\omega = f_\mu$ a.e./ω.

If $f, g \in \mathfrak{M}^*$ we define a linear functional fg on $\mathfrak{M}$ by

$$(fg, \nu) = \int f_\nu g_\nu \, d\nu.$$

Note that we also have $(fg, \nu) = \int f_\mu g_\mu \, d\nu$ whenever $\nu << \mu$, since $f_\mu = f_\nu$ a.e./ν and $g_\mu = g_\nu$ a.e./ν. For any pair $\nu_1, \nu_2 \in \mathfrak{M}$ we have $\nu_i << |\nu_1| + |\nu_2|$. Since $(fg, \nu_i) = \int f_\mu g_\mu \, d\nu_i$ for $\mu = |\nu_1| + |\nu_2|$, it follows that fg is linear. Furthermore,

$$|(fg, \nu)| \leqslant \|f_\nu g_\nu\| \|\nu\| = \|f_\nu\| \|g_\nu\| \|\nu\| \leqslant \|f\| \|g\| \|\nu\|.$$

Hence, fg is bounded and has norm dominated by $\|f\| \|g\|$.

The natural conjugation in $\mathfrak{M}^*$ defined by $(\bar{f}, \nu) = \overline{(f, \bar{\nu})}$ is obviously an involution on $\mathfrak{M}^*$ with the property that $\bar{f}_\mu = (f_\mu)^-$ for $\mu \in \mathfrak{M}$. To complete the proof that $\mathfrak{M}^*$ is a C^*-algebra, we must prove that $\|\bar{f}f\| = \|f\|^2$. Obviously, $\|\bar{f}\| = \|f\|$; and we have shown that $\|\bar{f}f\| \leqslant \|\bar{f}\|\,\|f\|$. Hence, let $\{\mu_n\}_{n=1}^\infty$ be a sequence of measures in $\mathfrak{M}$, of norm one, such that $\lim_n |(f, \mu_n)| = \|f\|$. If $\mu = \Sigma 2^{-n}|\mu_n|$ then $\mu_n \in L(\mu)$ for each n and, hence, $\|f_\mu\| = \|f\|$. Thus, $\|\bar{f}f\| \geqslant \|\bar{f}_\mu f_\mu\| = \|f_\mu\|^2 = \|f\|$. We conclude that $\|\bar{f}f\| = \|f\|^2$ and $\mathfrak{M}^*$ is a commutative C^*-algebra.

In a concrete representation of $\mathfrak{M}$ in $M(X, \Sigma)$, the functional 1, defined by $(1, \mu) = \mu(X)$, has the property that 1_μ is the identity of $L^\infty(\mu)$ for each μ. It follows that 1 is independent of the representation and is an identity for $\mathfrak{M}^*$.

§2.4. The standard representation

2.4.1. If A is a commutative C^*-algebra, X is its spectrum, and $a \to \hat{a}$ its Gelfand transform, then $a \to \hat{a}$ is a conjugation preserving isomorphism-isometry of A onto $C_0(X)$. If A has an identity, then X is compact and $a \to \hat{a}$ maps A onto $C(X)$ (cf. Gamelin [1]).

If $\mathfrak{M}$ is an L-space and X the spectrum of the C^*-algebra $\mathfrak{M}^*$, then we shall call X the standard domain of $\mathfrak{M}$ and identify $\mathfrak{M}^*$ with $C(X)$. Since $C(X)^* = M(X)$, there is an isometry $\mu \to \mu_X$ of $\mathfrak{M}$ into its second dual $\mathfrak{M}^{**} = M(X)$, which is characterized by

$$(f, \mu) = \int f\, d\mu_X$$

for $f \in C(X) = \mathfrak{M}^*$ and $\mu \in \mathfrak{M}$.

The map $\mu \to \mu_X$ is order preserving. In fact, if $\mu \in \mathfrak{M}$ then $\mu \geqslant 0$ if and only if $\int |f|^2\, d\mu_X = (\bar{f}f, \mu) \geqslant 0$ for each $f \in C(X)$, that is, if and only if $\mu_X \geqslant 0$.

The image $\mathfrak{M}_X$ of $\mathfrak{M}$ in $M(X)$ is an L-subspace of $M(X)$. This follows from the fact that for $\mu \in \mathfrak{M}$ each $\nu << \mu_X$ is a limit of measures of the form $f\mu_X = (f_\mu \mu)_X \in \mathfrak{M}_X$ for $f \in \mathfrak{M}^* = C(X)$. Thus, we have

2.4.2. THEOREM. *If M is an L-space, then there is a compact Hausdorff space X and a representation $\mu \to \mu_X$ of $\mathfrak{M}$ as an L-subspace of $M(X)$, such that $C(X)$ is identified with $\mathfrak{M}^*$ by the pairing $(f, \mu) = \int f\, d\mu_X$.*

We shall call X the standard domain and $\mu \to \mu_X$ the standard representation of $\mathfrak{M}$.

2.4.3. SUBALGEBRAS OF $\mathfrak{M}^*$. If A is any closed subalgebra of $\mathfrak{M}^*$ which is closed under involution and contains the identity, then A is also a commutative C^*-algebra and is isomorphic to $C(Y)$, where Y is its spectrum. The restriction map from the spectrum of $\mathfrak{M}^*$ to the spectrum of A is a continuous surjective map $\alpha : X \to Y$, with $f \to f \circ \alpha : C(Y) \to C(X) = \mathfrak{M}^*$ the isomorphism of $C(Y)$ onto A. It expresses Y as the factor space of X modulo the equivalence relation: $x \sim y$ if and only if $\alpha(x) = \alpha(y)$, i.e., if and only if $f(x) = f(y)$ for every $f \in A$.

The map $\mu \to \mu \circ \alpha^{-1} : M(X) \to M(Y)$ is an L-homomorphism. When composed with $\mu \to \mu_X$ it yields an L-homomorphism $\mu \to \mu_Y : \mathfrak{M} \to M(Y)$ characterized by

$f \circ \alpha, \mu) = \int f \, d\mu_Y$ for $f \in C(Y)$. Since $\mathfrak{M}$ separates points in A, the map $\mu \to \mu_Y$ must have weak-* dense range in $M(Y)$.

2.4.4. L-HOMOMORPHISMS. The appropriate class of morphisms in the category of L-spaces is the class of L-homomorphisms. This is because they are exactly those maps whose duals are C^*-algebra homomorphisms which preserve the identity. In fact,

PROPOSITION. *If* $\mathfrak{M}_1$ *and* $\mathfrak{M}_2$ *are* L*-spaces with standard domains* X_1 *and* X_2, *respectively, and if* $\phi : \mathfrak{M}_1 \to \mathfrak{M}_2$ *is a bounded linear map, then the following statements are equivalent*:

(a) ϕ *is an* L*-homomorphism*;

(b) $\phi^* : \mathfrak{M}_2^* \to \mathfrak{M}_1^*$ *is a* C^**-algebra homomorphism which maps the identy to the identity*:

(c) *there is a continuous map* $\alpha : X_1 \to X_2$ *with* $\phi\mu = \mu \circ \alpha^{-1}$ (*where* $\mathfrak{M}_1$ *and* $\mathfrak{M}_2$ *are identified with their images under the standard representations*).

PROOF. That (a) implies (b) follows directly from 2.2.5. If (b) holds then the restriction of the adjoint of ϕ^* to the spectrum of $\mathfrak{M}_1^* = C(X_1)$ yields the map α of (c). That (c) implies (a) follows from 2.2.2.

2.4.5. COROLLARY. *If* $\phi : \mathfrak{M}_1 \to \mathfrak{M}_2$ *is an* L*-homomorphism, then its image is an* L*-subspace of* $\mathfrak{M}_2$. *If* ϕ *is injective, it is an order preserving isometry.*

PROOF. If $\alpha : X_1 \to X_2$ is the map of 2.4.4(c), and $\nu << \phi\mu = \mu \circ \alpha^{-1}$ for some $\mu \in \mathfrak{M}_1$, then $\nu = f(\mu \circ \alpha^{-1})$ for some $f \in L^1(\mu \circ \alpha^{-1})$. It follows that $\nu = \omega \circ \alpha^{-1}$ for $\omega = (f \circ \alpha)\mu \in \mathfrak{M}_1$. Thus, the image of ϕ is an L-subspace of $\mathfrak{M}_2$.

If ϕ is injective, then $\phi^* : C(X_2) \to C(X_1)$ has dense range. However, since $\phi^* f = f \circ \alpha$, this forces α to be injective, ϕ^* to be surjective, and ϕ to be an isometry.

2.4.6. TOPOLOGY OF THE STANDARD DOMAIN. The standard domain X of an L-space $\mathfrak{M}$ is extremely disconnected; that is, each open set has open closure. In fact, if $U \subset X$ is open, then $\mu \to \int \chi_U \, d\mu$ is a bounded linear functional on $\mathfrak{M}$ and, hence, agrees with $\mu \to \int f \, d\mu$ for some $f \in C(X)$. Since $\mathfrak{M}$ is weak-* dense in $M(X)$, it follows easily that f must be the characteristic function of $\bar{U}$, the closure of U. This forces $\bar{U}$ to be open.

§2.5. Tensor products of L-spaces

2.5.1. Given two Banach spaces A and B, the greatest cross norm on the algebraic tensor product $A \otimes B$ is defined by

$$\|c\| = \inf\left\{\sum \|a_i\| \, \|b_i\| : c = \sum a_i \otimes b_i\right\}.$$

This norm clearly dominates every norm $\| \;\|'$ on $A \otimes B$ for which $\|a \otimes b\|' \leqslant \|a\| \, \|b\|$ for all $a \in A$, $b \in B$.

We denote the completion of $A \otimes B$ in the greatest cross norm by $A \hat{\otimes} B$. This Banach space has the property that if $\theta : A \times B \to C$ is a bilinear map of $A \times B$ into a Banach space C, and if $\|\theta(a, b)\| \leqslant K\|a\| \, \|b\|$ for a constant K, then the map $\phi : A \otimes B \to C$ defined by

$$\phi\left(\sum a_i \otimes b_i\right) = \sum \theta(a_i, b_i)$$

extends uniquely to a bounded linear map $\phi : A \hat{\otimes} B \to C$ with $\|\phi\| \leqslant K$ (Schatten [1]).

The object of this section is to characterize $\mathfrak{M} \hat{\otimes} \mathfrak{N}$ for L-spaces $\mathfrak{M}$ and $\mathfrak{N}$. The result is essentially that of Grothendieck [1].

2.5.2. PROPOSITION. *If $\mathfrak{M}$ and $\mathfrak{N}$ are L-spaces, then there is an order relation on $\mathfrak{M} \hat{\otimes} \mathfrak{N}$ so that it is also an L-space and $\mu \otimes \nu \geqslant 0$ whenever $\mu \geqslant 0$ and $\nu \geqslant 0$. Furthermore, if $\mathfrak{M}$ and $\mathfrak{N}$ are represented as L-subspaces of measure spaces $M(X, \Sigma)$ and $M(Y, \Sigma')$, then $\mathfrak{M} \hat{\otimes} \mathfrak{N}$ may be represented as the closed linear span in $M(X \times Y, \Sigma \times \Sigma')$ of the product measures $\mu \times \nu$ for $\mu \in \mathfrak{M}, \nu \in \mathfrak{N}$.*

PROOF. We define the positive cone in $\mathfrak{M} \hat{\otimes} \mathfrak{N}$ to be the closed convex hull of the set of measures of the form $\mu \otimes \nu$ for $\mu \geqslant 0, \nu \geqslant 0, \mu \in \mathfrak{M}, \nu \in \mathfrak{N}$. In order to prove that $\mathfrak{M} \hat{\otimes} \mathfrak{N}$ is an L-space with this order relation, it suffices to prove the second statement of the proposition, i.e., that there is an order preserving isometry of $\mathfrak{M} \hat{\otimes} \mathfrak{N}$ onto the L-subspace of $M(X \times Y, \Sigma \times \Sigma')$ generated by product measures $\mu \times \nu$.

Now the bilinear map $(\mu, \nu) \to \mu \times \nu : \mathfrak{M} \times \mathfrak{N} \to M(X \times Y, \Sigma \times \Sigma')$ satisfies $\|\mu \times \nu\| = \|\mu\| \|\nu\|$ and, hence, induces a norm decreasing linear map of $\mathfrak{M} \hat{\otimes} \mathfrak{N}$ into $M(X \times Y, \Sigma \times \Sigma')$. The range of this map is dense in the closed linear span of the measures $\mu \times \nu$ for $\mu \in \mathfrak{M}, \nu \in \mathfrak{N}$. Note that this map also preserves positivity. We shall prove that it is an isometry.

Given finitely many measures $\mu_1, \cdots, \mu_n \in \mathfrak{M}$, if we set $\mu = \Sigma |\mu_i|$, then there are functions $f_1, \cdots, f_n \in L^1(\mu)$ with $\mu_i = f_i \mu$ for each i. Furthermore, given $\epsilon > 0$, we may choose a disjoint partition $\{E_j\}_{j=1}^m \subset \Sigma$ of X such that each f_i is within ϵ, in $L^1(\mu)$-norm, of some linear combination of the functions χ_{E_j}. If we set $\rho_j = \chi_{E_j}\mu$ for $j = 1, \cdots, m$, then each μ_i is within ϵ, in norm, of some linear combination of the measures $\{\rho_j\}_{j=1}^m$. Now for $\nu_1, \cdots, \nu_n \in \mathfrak{N}$ it follows that the element $\Sigma \mu_i \otimes \nu_i \in \mathfrak{M} \otimes \mathfrak{N}$ may be approximated to within $\epsilon \Sigma \|\nu_i\|$ by an element of the form $\Sigma \rho_j \otimes \omega_j$, with each ω_j an appropriate linear combination of the measures ν_i.

We conclude from the above that elements of the form $\Sigma \rho_j \otimes \omega_j$, for which the ρ_j's have disjoint supports in X, are dense in $\mathfrak{M} \hat{\otimes} \mathfrak{N}$. For such an element, we have that the measures $\rho_j \times \omega_j$ have disjoint supports in $X \times Y$ and, hence, that

$$\left\|\sum \rho_j \times \omega_j\right\| = \sum \|\rho_j \times \omega_j\| = \sum \|\rho_j\| \|\omega_j\| \geqslant \left\|\sum \rho_i \otimes \omega_j\right\|.$$

Since we already have that $\|\Sigma \rho_j \times \omega_j\| \leqslant \|\Sigma \rho_j \otimes \omega_j\|$, we conclude that equality holds and our map of $\mathfrak{M} \hat{\otimes} \mathfrak{N}$ into $M(X \times Y, \Sigma \times \Sigma')$ is an isometry.

To complete the proof, we must verify that the closed linear span $\langle \mathfrak{M} \times \mathfrak{N} \rangle$ of $\{\mu \times \nu : \mu \in \mathfrak{M}, \nu \in \mathfrak{N}\}$ is indeed an L-subspace of $M(X \times Y, \Sigma \times \Sigma')$. First, note that for $\mu \in \mathfrak{M}, \nu \in \mathfrak{N}$, the space $L(\mu \times \nu)$ is contained in $\langle \mathfrak{M} \times \mathfrak{N} \rangle$. This follows from the fact that each $f \in L^1(\mu \times \nu)$ is approximable, in $L^1(\mu \times \nu)$-norm, by linear combinations of functions $\chi_{E \times F}$ for $E \in \Sigma, F \in \Sigma'$. Now if $\{\mu_i\}_{i=1}^\infty$ and $\{\nu_i\}_{i=1}^\infty$ are countable sets in $\mathfrak{M}$ and $\mathfrak{N}$ and we set $\mu = \Sigma 2^{-n} \|\mu_n\|^{-1} |\mu_n|, \nu = \Sigma 2^{-n} \|\nu_n\|^{-1} |\nu_n|$, then any linear combination of product measures $\mu_i \times \nu_i$ is in $L(\mu \times \nu)$. It follows that each element of $\overline{\langle \mathfrak{M} \times \mathfrak{N} \rangle}$ is in $L(\mu \times \nu)$ for some $\mu \in \mathfrak{M}, \nu \in \mathfrak{N}$, we conclude that $\overline{\langle \mathfrak{M} \times \mathfrak{N} \rangle}$ is an L-subspace.

2.5.3. REMARK. Given the standard representations $\mu \to \mu_X$ and $\nu \to \nu_Y$ of $\mathfrak{M}$ and $\mathfrak{N}$, one might be tempted to conclude that the corresponding representation (determined by $\mu \otimes \nu \to \mu_X \times \nu_Y$) of $\mathfrak{M} \hat{\otimes} \mathfrak{N}$ in $M(X \times Y)$ is the standard representation. However, this is true if and only if either $\mathfrak{M}$ or $\mathfrak{N}$ is finite dimensional (X or Y is a finite set). Otherwise, it is easy to see that $X \times Y$ is not extremely disconnected. In fact, if $\{U_i\}_{i=1}^{\infty}$ and $\{V_i\}_{i=1}^{\infty}$ are disjoint sequences of open closed sets in X and Y, respectively, then the set $W = \bigcup_{i=1}^{\infty} U_i \times V_i$ is open in $X \times Y$, but its closure is not; if it were, it would be the union of finitely many open-closed rectangles of the form $U \times V$, and it is not difficult to see that this is impossible.

CHAPTER 3. CONVOLUTION MEASURE ALGEBRAS

In this chapter we define the class of convolution measure algebras. It is this class to which the techniques of succeeding chapters are applicable. If $M(S)$ is the measure algebra of a locally compact semigroup, then any subalgebra of $M(S)$ which is also an L-subspace is a convolution measure algebra. Important examples are $M(G)$ and $L(G)$ for an l. c. a. group G.

§3.1. **Definition and examples**

3.1.1. Given any Banach algebra A, the multiplication map $(a, b) \to ab : A \times A \to A$ is a bilinear map satisfying $\|ab\| \leqslant \|a\| \|b\|$. It follows that there is a unique norm decreasing linear map $\pi : A \hat{\otimes} A \to A$ for which $\pi(a \otimes b) = a \cdot b$ (2.6.1).

3.1.2. Definition. A convolution measure algebra is a Banach algebra $\mathfrak{M}$ with an order relation under which $\mathfrak{M}$ is an L-space and the product map $\pi : \mathfrak{M} \hat{\otimes} \mathfrak{M} \to \mathfrak{M}$ is an L-homomorphism.

An L-subalgebra of a convolution measure algebra is a subalgebra which is also an L-subspace. Similarly, an L-ideal is an ideal which is also an L-subspace.

Note that an L-subalgebra of a convolution measure algebra is also a convolution measure algebra.

3.1.3. Proposition. *If S is a locally compact topological semigroup, then $M(S)$ (and, hence, each of its L-subalgebras) is a convolution measure algebra.*

Proof. It follows from 2.6.2 that $M(S) \hat{\otimes} M(S)$ may be represented as the closed linear span in $M(S \times S)$ of the product measures $\mu \times \nu$.

Now for $\mu, \nu \in M(S)$, the convolution product $\mu * \nu$ can be described as the image of $\mu \times \nu$ under the map $\omega \to \omega \circ \alpha^{-1} : M(S \times S) \to M(S)$ induced by $\alpha : S \times S \to S$, where $\alpha(s, t) = st$. In fact, for $f \in C_0(S)$,

$$\int f \, d(\mu \times \nu) \circ \alpha^{-1} = \int (f \circ \alpha) \, d(\mu \times \nu) = \iint f(st) \, d\mu(s) \, d\nu(t) = \int f \, d(\mu * \nu).$$

It follows that the map $\pi : M(S) \hat{\otimes} M(S) \to M(S)$, induced by convolution multiplication, is just the L-homomorphism $\mu \to \mu \circ \alpha^{-1} : M(S \times S) \to M(S)$ restricted to $M(S) \hat{\otimes} M(S)$. Since this is an L-homomorphism, $M(S)$ is a convolution measure algebra.

We should remark that it is only necessary to assume separate continuity of multiplication on S in order to conclude that $M(S)$ is a convolution measure algebra. However, the proof is more difficult (cf. Taylor [6]).

The main theorem (3.2.3) of this chapter is a partial converse of the above result.

§3.2. The structure semigroup

3.2.1. COMPLEX HOMOMORPHISMS. Given elements $f, g \in \mathfrak{M}^*$, for an L-space $\mathfrak{M}$, we define an element $f \otimes g \in (\mathfrak{M} \hat{\otimes} \mathfrak{M})^*$ to be the continuous linear extension to $\mathfrak{M} \hat{\otimes} \mathfrak{M}$ of the bilinear form $(\mu, \nu) \to (f, \mu)(g, \nu)$.

Note that if $\mathfrak{M}$ is identified with its image in $M(X)$ under the standard representation, $\mathfrak{M}^*$ with $C(X)$, and $\mathfrak{M} \hat{\otimes} \mathfrak{M}$ with its image in $M(X \times X)$ under its corresponding representation (as in 2.7.2), then the element $f \otimes g \in (\mathfrak{M} \hat{\otimes} \mathfrak{M})^*$ can be identified with the function $(x, y) \to f(x)g(y)$ on $X \times X$.

PROPOSITION. *If $\mathfrak{M}$ is a commutative convolution measure algebra and $0 \neq f \in \mathfrak{M}^*$, then f is a complex homomorphism of $\mathfrak{M}$ if and only if $\pi^* f = f \otimes f$, where $\pi^* : \mathfrak{M}^* \to (\mathfrak{M} \otimes \mathfrak{M})^*$ is the adjoint of the product map $\pi : \mathfrak{M} \hat{\otimes} \mathfrak{M} \to \mathfrak{M}$.*

PROOF. We have that f is a complex homomorphism if and only if $(f, \mu * \nu) = (f, \mu)(f, \nu)$ for $\mu, \nu \in \mathfrak{M}$, that is, if and only if

$$(\pi^* f, \mu \otimes \nu) = (f, \pi(\mu \otimes \nu)) = (f, \mu * \nu) = (f, \mu)(f, \nu) = (f \otimes f, \mu \otimes \nu).$$

Since the span of the measures $\mu \otimes \nu$ is dense in $\mathfrak{M} \hat{\otimes} \mathfrak{M}$, the proposition follows.

3.2.2. PROPOSITION. *Let $\mathfrak{M}$ be a commutative convolution measure algebra, $\Delta \subset \mathfrak{M}^*$ its spectrum, and A the closed linear span of Δ in $\mathfrak{M}^*$. Then A is a C^*-subalgebra of $\mathfrak{M}^*$ containing the identity and $\pi^* A$ is contained in $A \bar{\otimes} A$, the closed linear span in $(\mathfrak{M} \hat{\otimes} \mathfrak{M})^*$ of the elements $f \otimes g$ for $f, g \in A$.*

The proposition follows immediately from 3.2.1 and the following

LEMMA. *The set $\Delta \cup \{0\} \subset \mathfrak{M}^*$ contains the identity functional and is closed under products and conjugation.*

PROOF. For the standard representation, we identify $\mathfrak{M}$ with its image in $M(X)$, $\mathfrak{M}^*$ with $C(X)$, and $\mathfrak{M} \hat{\otimes} \mathfrak{M}$ with its image in $M(X \times X)$. Then, by 2.4.4, for $f, g \in \Delta \subset C(X)$ we have

$$(\pi^* \bar{f})(x, y) = \overline{(\pi^* f)}(x, y) = \overline{f(x) f(y)} = (\bar{f} \otimes \bar{f})(x, y)$$

and

$$\pi^*(fg)(x, y) = (\pi^* f)(\pi^* g)(x, y) = f(x) f(y) g(x) g(y)$$
$$= (fg \otimes fg)(x, y).$$

Hence, $\bar{f}$ and fg are elements of Δ (unless $fg = 0$, in which case $fg \in \Delta \cup \{0\}$). Since $\pi^* 1$ is the identity, $1 \otimes 1$, of $(\mathfrak{M} \hat{\otimes} \mathfrak{M})^*$, we have $1 \in \Delta$.

3.2.3. THEOREM. *If $\mathfrak{M}$ is a commutative convolution measure algebra, then there is a compact, abelian topological semigroup S and an algebra homomorphism and L-homomorphism $\mu \to \mu_S : \mathfrak{M} \to M(S)$ such that*

(a) *each complex homomorphism of $\mathfrak{M}$ is determined by a semicharacter $f \in \hat{S}$*

through the pairing $(f, \mu) = \int f\, d\mu_S$,

(b) $\hat{S}$ *separates points in* S, *and*

(c) $\mu \to \mu_S$ *has weak-$*$ dense image in* $M(S)$.

PROOF. Let $\mathfrak{M} \to M(X)$ be the standard representation and identify $\mathfrak{M}$ with its image. Let A be the closed linear span of the spectrum Δ of $\mathfrak{M}$. Since A is a C^*-subalgebra of $\mathfrak{M}^*$ containing the identity 3.2.2 , there is a compact Hausdorff space S and a continuous surjection $\alpha : X \to S$, such that $f \to f \circ \alpha$ is an isomorphism of $C(S)$ onto $A \subset \mathfrak{M}^* = C(X)$. The map $\mu \to \mu \circ \alpha^{-1} : M(X) \to M(S)$ is an L-homomorphism which maps $\mathfrak{M}$ onto a weak-$*$ dense subspace of $M(S)$. For $\mu \in \mathfrak{M}$ we set $\mu_S = \mu \circ \alpha^{-1}$.

The semigroup operation on S is defined as follows: by 3.2.2, $f \in A$ implies $\pi^* f$ is in the closed linear span $A \mathbin{\overline{\otimes}} A$ in $(\mathfrak{M} \mathbin{\hat{\otimes}} \mathfrak{M})^*$ of elements of the form $f \otimes g$ for $f, g \in A$. Hence, for $f \in A$, $\pi^* f$ can be represented as a continuous function on $X \times X$ with the property that $\pi^* f(x, y) = \pi^* f(x', y')$ whenever $\alpha(x) = \alpha(x')$ and $\alpha(y) = \alpha(y')$. It follows that $\pi^* f$ determines a continuous function θf on $S \times S$ such that $\theta f(\alpha(x), \alpha(y)) = \pi^* f(x, y)$. Note that

$$\int_X f\, d\mu * \nu = \int_{X \times X} \pi^* f\, d(\mu \times \nu) = \int_{S \times S} \theta f\, d(\mu_S \times \nu_S)$$

for $\mu, \nu \in \mathfrak{M}$. The composition $g \to g \circ \alpha \to \theta(g \circ \alpha)$ is then a homomorphism N of $C(S)$ into $C(S \times S)$ such that $(Ng)(\alpha(x), \alpha(y)) = \pi^*(g \circ \alpha)(x, y)$. Note that $N1 = 1$ and

(i) $$\int_S g\, d(\mu * \nu)_S = \int_{S \times S} Ng\, d(\mu_S \times \nu_S).$$

It follows that there is a continuous map $S \times S \to S$, which we denote by $(s, t) \to st$, such that $Ng(s, t) = g(st)$ for $g \in C(S)$, $s, t \in S$. Equation (i) then becomes

(ii) $$\int_S g\, d(\mu * \nu)_S = \int_S \int_S g(st)\, d\mu_S(s)\, d\nu_S(t).$$

Equation (ii), together with the associativity and commutativity of $(\mu, \nu) \to \mu * \nu$ imply that the operation $(s, t) \to st$ is also associative and commutative. Hence, S is an abelian, compact topological semigroup. Equation (ii) also implies that $(\mu * \nu)_S = \mu_S * \nu_S$, so that $\mu \to \mu_S$ is an algebra homomorphism of $\mathfrak{M}$ into the algebra $M(S)$.

Now for the isomorphism $f \to f \circ \alpha : C(S) \to A$, we have that $f \circ \alpha \in \Delta$ if and only if $f \in \hat{S}$. In fact, the identity $f(st) = f(s)f(t)$ means exactly that for $x, y \in X$,

$$\begin{aligned} \pi^*(f \circ \alpha)(x, y) &= (Nf)(\alpha(x), \alpha(y)) = f(\alpha(x)\alpha(y)) \\ &= f(\alpha(x)) f(\alpha(y)) = [(f \circ \alpha) \otimes (f \circ \alpha)](x, y). \end{aligned}$$

Part (a) of the theorem follows immediately. Part (b) then follows from the fact that A is the closed linear span of Δ. We have already observed that (c) is a direct consequence of the construction of the map $\mu \to \mu_S$.

3.2.4. COROLLARY. *If* $\mathfrak{M}$ *is a commutative convolution measure algebra, then so is* $\mathfrak{M}/\mathrm{Rad}(\mathfrak{M})$, *where* $\mathrm{Rad}(\mathfrak{M})$ *is the intersection of the regular maximal ideals of* $\mathfrak{M}$. *For a semisimple convolution measure algebra* $(\mathrm{Rad}(\mathfrak{M}) = 0)$, *the map* $\mu \to \mu_S$ *of* 3.2.3 *is an isometry.*

PROOF. If $\mathfrak{M}_S$ is the image of $\mu \to \mu_S : \mathfrak{M} \to M(S)$, then $\mathfrak{M}_S$ is an L-subalgebra of $M(S)$ by 2.4.5. The kernel of $\mu \to \mu_S$ is $\operatorname{Rad}(\mathfrak{M})$, and it follows readily that it induces an isomorphism-isometry of $\mathfrak{M}/\operatorname{Rad}(\mathfrak{M})$ onto $\mathfrak{M}_S$.

3.2.5. TERMINOLOGY. The semigroup A, constructed above, will be called the structure semigroup of $\mathfrak{M}$. The uniqueness of S and $\mu \to \mu_S$ (to within the obvious equivalence) subject to the conditions (a), (b), and (c) of 3.2.3, is fairly apparent; in any case, it follows from the results of the next section.

Henceforth, we shall identify $\hat{S}$ with the spectrum Δ of $\mathfrak{M}$. With this identification, the Gelfand transform $\mu \to \hat{\mu} : \mathfrak{M} \to C_0(\hat{S})$ is given by

$$\hat{\mu}(f) = \int f \, d\mu_S.$$

The (Gelfand or weak) topology for $\hat{S}$ is the weakest topology for which each $\hat{\mu}$ for $\mu \in \mathfrak{M}$ is continuous. This is the weak-$*$ topology inherited from $\mathfrak{M}^*$ by the identification of $\hat{S}$ with $\Delta \subset \mathfrak{M}^*$.

§3.3. Homomorphisms of measure algebras

3.3.1. The term "convolution measure algebra" is somewhat clumsy. In the remainder of the book we shall shorten it to "measure algebra".

In this section we shall study the appropriate class of "morphisms" for measure algebras.

DEFINITION. If $\mathfrak{M}$ and $\mathfrak{N}$ are measure algebras, then a homomorphism of measure algebras, from $\mathfrak{M}$ to $\mathfrak{N}$, will be an algebra homomorphism which is also an L-homomorphism. An isomorphism of measure algebras is an invertible homomorphism of measure algebras. Note that, by 2.4.5, such a thing must be an order preserving isometry.

Recall that the map $\mu \to \mu_S : \mathfrak{M} \to M(S)$ of 3.2.3 is a homomorphism of measure algebras.

3.3.2. SEMIGROUP HOMOMORPHISMS. Note that a continuous semigroup homomorphism $\alpha : S \to T$, between locally compact semigroups, induces a map $\mu \to \mu \circ \alpha^{-1} : M(S) \to M(T)$ which is clearly a homomorphism of measure algebras, as is its restriction to any L-subalgebra of $M(S)$. Conversely, we have

3.3.3. PROPOSITION. *If $\mathfrak{M}$ and $\mathfrak{N}$ are commutative measure algebras and $\phi : \mathfrak{M} \to \mathfrak{N}$ is a homomorphism of measure algebras, then there is a continuous semigroup homomorphism $\alpha : S \to T$, between the structure semigroups of $\mathfrak{M}$ and $\mathfrak{N}$, such that the diagram*

(i)
$$\begin{array}{ccc} \mathfrak{M} & \xrightarrow{\ \phi\ } & \mathfrak{N} \\ {\scriptstyle \mu \to \mu_S} \downarrow & & \downarrow {\scriptstyle \nu \to \nu_T} \\ M(S) & \xrightarrow{\ \mu \to \mu \circ \alpha^{-1}\ } & M(T) \end{array}$$

commutes.

PROOF. Since ϕ is an algebra homomorphism, its adjoint $\phi^* : \mathfrak{N}^* \to \mathfrak{M}^*$ maps $\Delta(\mathfrak{N}) \cup \{0\}$ into $\Delta(\mathfrak{M}) \cup \{0\}$, where $\Delta(\mathfrak{M})$ and $\Delta(\mathfrak{N})$ are the spectra of $\mathfrak{M}$ and $\mathfrak{N}$, respectively. If A and B are the closed linear spans of $\Delta(\mathfrak{M})$ and $\Delta(\mathfrak{N})$, then (since ϕ is an L-homomorphism) ϕ^* is an algebra homomorphism of B into A with $\phi^*1 = 1$.

Since $A \simeq C(S)$ and $B \simeq C(T)$, it follows that there is a continuous map $\alpha : S \to T$ for which (i) is commutative.

To see that α is a semigroup homomorphism, note that $f \to f \circ \alpha : C(T) \to C(S)$ maps $\hat{T} \cup \{0\}$ into $\hat{S} \cup \{0\}$ (since $\phi^* : B \to A$ maps $\Delta(\mathfrak{N}) \cup \{0\}$ into $\Delta(\mathfrak{M}) \cup \{0\}$). Thus, for $f \in \hat{T}$, $s, t \in S$ we have

$$f(\alpha(st)) = f \circ \alpha(st) = (f \circ \alpha)(s)(f \circ \alpha)(t) = f(\alpha(s))f(\alpha(t)) = f(\alpha(s)\alpha(t)).$$

Since $\hat{T}$ separates points in T we conclude that $\alpha(st) = \alpha(s)\alpha(t)$ and α is a semigroup homomorphism,

3.3.4. It is often useful to know that the map $\alpha : S \to T$ of 3.3.3 is injective. Thus, with $\phi : \mathfrak{M} \to \mathfrak{N}$ and $\alpha : S \to T$ as in 3.3.3, we prove

PROPOSITION. *The map $\alpha : S \to T$ is injective and every continuous semicharacter on $\alpha(S)$ extends to an element of $\hat{T}$, if and only if each complex homomorphism of $\mathfrak{M}$ is the image of a complex homomorphism of $\mathfrak{N}$ under $\phi^* : \mathfrak{N}^* \to \mathfrak{M}^*$.*

PROOF. Clearly $\Delta(\mathfrak{M}) \subset \phi^*\Delta(\mathfrak{N})$ if and only if each $f \in \hat{S}$ has the form $g \circ \alpha$ for some $g \in \hat{T}$. Since $\hat{S}$ separates points in S, this is equivalent to the fact that α is injective and each semicharacter on its image extends to an element of $\hat{T}$.

§3.4. Elementary properties of S and $\hat{S}$

3.4.1. IDENTITIES. If A is a commutative Banach algebra, then a weak approximate identity for A is a net $\{a_\alpha\} \subset A$ such that the net of Gelfand transforms $\{\hat{a}_\alpha\}$ converges pointwise to 1 on the spectrum of A.

PROPOSITION. *If S is the structure semigroup of a commutative measure algebra $\mathfrak{M}$, then S has an identity e if and only if there is a weak approximate identity $\{\nu_\alpha\}$ for $\mathfrak{M}$ with $\|\nu_\alpha\| = 1$ for each α. In this case, $(\nu_\alpha)_S \to \delta_e$ in the weak-* topology of $M(S)$.*

PROOF. If $\{\nu_\alpha\}$ is a weak approximate identity with $\|\nu_\alpha\| = 1$, then the net $\{(\nu_\alpha)_S\}$ has a weak-* cluster point $\nu \in M(S)$, with $\|\nu\| \leqslant 1$. For any such cluster point, we must have

$$\int f\, d\nu = \lim_\alpha \hat{\nu}_\alpha(f) = 1 \quad \text{for all } f \in \hat{S}.$$

It follows that for all $\mu \in M(S)$ and $f \in \hat{S}$,

$$\int f\, d\mu * \nu = \int f\, d\mu \int f\, d\nu = \int f\, d\mu.$$

Since the linear span of $\hat{S}$ is dense in $C(S)$, we conclude that $\mu * \nu = \mu$ for all $\mu \in M(S)$; that is, ν is an identity for $M(S)$.

Now if $s \in S$, then $\delta_s * \nu = \delta_s$ implies that $\nu(\{t : st = s\}) = 1$. Since $\|\nu\| \leqslant 1$, we conclude that ν is concentrated on the set $\{t \in S : st = s$ for all $s \in S\}$. This set can contain only one point–an identity e for S. Hence, $\nu = \delta_e$ and the net $\{(\nu_\alpha)_S\}$ converges in the weak-* topology to δ_e.

Conversely, if S has an identity e and $\{U_\alpha\}$ is a neighborhood base at e, we can choose for each α a measure $\nu_\alpha \in \mathfrak{M}$ with $\nu_\alpha \geqslant 0$, $\|\nu_\alpha\| = 1$, and $(\nu_\alpha)_S$ is concentrated on U_α. This follows from the fact that $\mathfrak{M}_S$ is weak-$*$ dense in $M(S)$. Clearly, $\{\nu_\alpha\}$ is a weak approximate identity for $\mathfrak{M}$.

3.4.2. COROLLARY. *If $\mathfrak{M}$ has an identity δ of norm one, then S has an identity e and $\delta_S = \delta_e$.*

For a measure algebra without an identity of norm one, it is an easy matter to adjoin one. In fact, if we give $\mathfrak{M} \oplus \mathbf{C}$ the norm "$\|\mu \oplus z\| = \|\mu\| + |z|$", product "$(\mu \oplus z)(\nu \oplus w) = (\mu\nu + z\nu + w\nu) \oplus zw$", and order relation "$(\mu \oplus z) \geqslant 0$ if and only if $\mu \geqslant 0, z \geqslant 0$", then it is easy to see that $\mathfrak{M} \oplus \mathbf{C}$ is a measure algebra with an identity $\delta = 0 \oplus 1$ of norm one. If S is the structure semigroup for $\mathfrak{M}$, the structure semigroup for $\mathfrak{M} \oplus \mathbf{C}$ is the discrete-union $S \cup \{e\}$ of S with a singleton set $\{e\}$ for which $e^2 = e$ and $es = s$ for all $s \in S$. The map $(\mu \oplus z) \to (\mu \oplus z)_{S \cup \{e\}}$ of 3.2.3 is defined by $(\mu \oplus z)_{S \cup \{e\}} = \mu_S + z\delta_e$.

3.4.3. OPERATIONS IN $\hat{S}$. If S is any topological semigroup, then $\hat{S}$ is closed under conjugation and (nonzero) products. It could happen that $fg = 0 \notin \hat{S}$ for some nonzero pair $f, g \in \hat{S}$. However, this is impossible if S has an identity e; in this case, $f(e) = 1$ for every $f \in \hat{S}$.

Thus, for a topological semigroup S with identity, $\hat{S}$ is a semigroup (under pointwise multiplication) with an involution (conjugation). Furthermore, if $f \in \hat{S}$ then $|f| \in \hat{S}$ and, in fact, $|f|^z \in \hat{S}$ for all $z \in \mathbf{C}$ with $\operatorname{Re} z > 0$. The semigroup $\hat{S}$ has an identity since $1 \in \hat{S}$.

In later chapters on the structure of $\hat{S}$, for S the structure semigroup of a measure algebra, these operations and the pointwise order relation, $f \leqslant g$, on the positive elements of $\hat{S}$ will be of great importance.

Unfortunately, for S the structure semigroup of $\mathfrak{M}$, the Gelfand topology on $\hat{S}$, induced by $\mathfrak{M}$, does not behave well with respect to the above operations. Although $(f, g) \to fg : \hat{S} \times \hat{S} \to \hat{S}$ is separately continuous and $f \to \bar{f}$ is continuous, $(f, g) \to fg$ is not generally jointly continuous and $f \to |f| : \hat{S} \to \hat{S}$ is not generally continuous. That conjugation is continuous and multiplication is separately continuous are easily verified using the fact that $f_\alpha \to f$ in the Gelfand topology if and only if $\int f_\alpha \, d\mu \to \int f \, d\mu$ for every $\mu \in \mathfrak{M}_S$. In the next chapter we will show that for $\mathfrak{M} = M(G)$ the multiplication in $\hat{S}$ fails to be jointly continuous and $f \to |f|$ fails to be continuous.

3.4.4. POLAR DECOMPOSITION. If S is the structure semigroup of a measure algebra $\mathfrak{M}$ and g is a bounded Borel semicharacter on S, then the pairing $(g, \mu) = \int g \, d\mu_S$ defines a complex homomorphism of $\mathfrak{M}$. It follows that there is an $f \in \hat{S}$ such that $\int g \, d\mu_S = \int f \, d\mu_S$ for every $\mu \in \mathfrak{M}$. Hence, $f = g$ a.e. with respect to every measure in $\mathfrak{M}_S$. This leads to a fact which is quite special to structure semigroups:

PROPOSITION. *If $f \in \hat{S}$ then there is a unique element $h \in \hat{S}$ with $f = h|f|$, $|h|^2 = |h|$, and $h = 0$ on the interior of the set where $f = 0$.*

PROOF. If g is defined by $g(s) = f(s)|f(s)|^{-1}$, if $f(s) \neq 0$ and $g(s) = 0$, if $f(s) = 0$, then g is a bounded Borel semicharacter. Hence, $g = h$ a.e. for some $h \in \hat{S}$, where the term "a.e." refers to equality almost everywhere with respect to each measure in $\mathfrak{M}_S$.

Note that $f = h|f|$ a.e. and, hence, equality holds everywhere since f and $h|f|$ are continuous and $\mathfrak{M}_S$ is weak-* dense in $M(S)$. Similarly, $|g|^2 = |g|$ implies that $|h|^2 = |h|$ a.e. and, hence, $|h|^2 = |h|$ everywhere; in other words, $|h|(S) \subset \{0, 1\}$.

Note that $\{s : h(s) = 0\}$ is open and closed and equal a.e. to the closed set $\{s : g(s) = 0\} = \{s : f(s) = 0\}$. Since no open set can have μ-measure zero for all $\mu \in \mathfrak{M}_S$, we conclude that $\{s : h(s) = 0\}$ is the interior of $\{s : f(s) = 0\}$.

Clearly, the equality $f = h|f|$ and the condition that h vanish on the interior of the set where f vanishes uniquely determine h.

Observe that if $f = h|f|$ is the polar decomposition of $f \in \hat{S}$ and if $|f|^2 \neq |f|$, then $z \to h|f|^z$ (Re $(z) \geqslant 0$) is a nonconstant analytic map and serves to embed an analytic disc about f in $\hat{S}$. This will be important in Chapter 10.

3.4.5. IDEALS OF S. For an abelian semigroup S, an ideal of S is a set J with $SJ \subset J$. The product J_1J_2, intersection $J_1 \cap J_2$, and union $J_1 \cup J_2$ of ideals are ideals and $J_1J_2 \subset J_1 \cap J_2$.

Note that if S is a compact abelian topological semigroup, then a principal ideal tS must be compact since $s \to ts$ is continuous. Hence, every ideal contains a compact ideal. Since compact ideals have the finite intersection property, we conclude that every compact abelian topological semigroup contains a unique minimal ideal K and that K is compact. The ideal K is called the kernel of S.

If K is a minimal ideal of any abelian semigroup S, then $sK = K$ for any $s \in K$. Hence, there is $p \in K$ such that $ps = s$. Similarly, there is an element $s' \in K$ such that $s's = p$. Note that $\{t \in K : pt = t\}$ is a nonempty subideal of K and, hence, must be equal to K. We conclude that K is a group with identity p.

Now a group G which is a compact Hausdorff space with a continuous multiplication must be a topological group; that is, inversion is also continuous. In fact, if $\{g_\alpha\}$ is a net in G with $g_\alpha \to g$ and g' is any cluster point of the net $\{g_\alpha^{-1}\}$, we have from continuity of multiplication that gg' is a cluster point of $\{g_\alpha g_\alpha^{-1}\}$; i.e., $gg' = e$ and $g' = g^{-1}$.

From the above considerations we conclude:

PROPOSITION. *If S is a compact abelian topological semigroup, then S contains a unique minimal ideal K (the kernel of S). Furthermore, the ideal K is a compact topological group.*

3.4.6. EXTENSION OF SEMICHARACTERS. If J is an ideal of an abelian topological semigroup S, then each element of $\hat{J}$ extends uniquely to an element of $\hat{S}$. In fact, if $g \in \hat{J}$ then $g(s) \neq 0$ for some $s \in J$. If we set $f(t) = g(ts)g(s)^{-1}$ for $t \in S$, then f is continuous, bounded, and

$$\begin{aligned} f(t_1t_2) &= g(t_1t_2s)g(s)^{-1} = g(t_1s\ t_2s)g(s)^{-2} \\ &= (g(t_1s)g(s)^{-1})\,(g(t_2s)g(s)^{-1}) = f(t_1)f(t_2); \end{aligned}$$

hence, $f \in \hat{S}$. Clearly $f|_J = g$ and the equation

$$f(t) = f(ts)f(s)^{-1} = g(ts)g(s)^{-1},$$

which must hold for any extension, uniquely defines f.

We conclude from the above that restriction $f \to f|_J$ maps $\{f \in \hat{S} : f(J) \neq (0)\}$ isomorphically onto $\hat{J}$.

If K is a minimal ideal of S, then elements of $\hat{K}$ are group characters and, hence, have modulus one. It follows that, for $f \in \hat{S}$, either $f(K) = 0$ or $|f| = 1$. Thus, restriction $f \to f|_K$ is an isomorphism of the subgroup $\{f \in \hat{S} : |f| = 1\}$ of $\hat{S}$ onto $\hat{K}$. Note that $\{f \in \hat{S} : |f| = 1\}$ is the group of elements of $\hat{S}$ which have inverses relative to the identity 1 of $\hat{S}$ ($\bar{f}$ is the inverse of f in this group).

3.4.7. GROUPS IN S. If S is any abelian semigroup and $p = p^2$ is an idempotent of S, then there is a unique maximal subgroup of S containing p. In fact if $G_p = \{s \in S : ps = s\} \cap \{t \in S : p \in tS\}$ then G_p is clearly a group in S with p as identity and G_p contains every other such group.

Note that, in case S is a compact topological semigroup, the maximal group G_p is closed. Certainly $\{s \in S : ps = s\}$ is closed. To see that $\{t \in S : p \in tS\}$ is closed, note that if $t_\alpha \to t$ and $s_\alpha \in S$ with $s_\alpha t_s = p$ for each α, then any cluster point s of $\{s_\alpha\}$ will satisfy $st = p$. Thus, a maximal group in a compact topological semigroup is a compact topological group.

In any topological semigroup the set of all idempotents is clearly closed. For a compact semigroup, the union of the corresponding maximal groups is also closed. In fact, if $s_\alpha \to s$ and $s_\alpha \in G_{p_\alpha}$, then it follows readily that $s \in G_p$ for any cluster point p of $\{p_\alpha\}$.

If S is the structure semigroup of a measure algebra $\mathfrak{M}$, the preceding considerations suggest the following questions: For which measures $\mu \in \mathfrak{M}$ is it true that μ_S is concentrated on a maximal group of S? For which μ is μ_S concentrated on the union of the maximal groups of S? The first of these questions turns out to be of paramount importance in succeeding chapters. A final answer is obtained in Chapter 7. The second question is, as yet, unanswered.

§3.5. L-ideals of $\mathfrak{M}$

3.5.1. In this section, for simplicity of discussion, we assume that $\mathfrak{M}$ is a semisimple commutative measure algebra with structure semigroup S, and we identify $\mathfrak{M}$ with its image $\mathfrak{M}_S$ in $M(S)$. Recall that an L-ideal of $\mathfrak{M}$ is an ideal $\mathfrak{N}$ which is also an L-subspace.

Note that if I is any ideal of a commutative Banach algebra A, then each complex homomorphism of I has a unique extension to a complex homomorphism of A. In fact, if $g \in \Delta(I)$, we simply define its unique extension $f \in \Delta(A)$ by $f(a) = g(ab)g(b)^{-1}$, where $b \in I$ is any element for which $g(b) \neq 0$. Thus, the restriction map $f \to f|_I$ is bijective from $\{f \in \Delta(A) : f(I) \neq (0)\}$ to $\Delta(I)$.

If we combine this fact with 3.3.4 we obtain

3.5.2. PROPOSITION. *If* $\mathfrak{N}$ *is an L-ideal of* $\mathfrak{M}$, *then the structure semigroup of* $\mathfrak{N}$ *is the ideal* J *of* S *consisting of the smallest closed set on which each element of* $\mathfrak{N}$ *is concentrated. We call this set the support of* $\mathfrak{N}$.

PROOF. By 3.3.4 and 3.5.1 the injection $\mathfrak{N} \to \mathfrak{M}$ induces an injection of the structure semigroup of $\mathfrak{N}$ into S. If J is the image of this injection, then every element of $\mathfrak{N}$ is concentrated on J and $\mathfrak{N}$ is weak-$*$ dense in $M(J)$. It follows that J is the smallest closed set in S supporting $\mathfrak{N}$.

Now the fact that $\mathfrak{M} * \mathfrak{N} \subset \mathfrak{N}$, on passing to weak-$*$ closures; implies that $\mathfrak{M} * M(J) \subset M(J)$ and then that $M(S) * M(J) \subset M(J)$ (we use here the fact that convolution is separately weak-$*$ continuous in $M(S)$). Applied to point measures, this implies that J is an ideal of S.

3.5.3. RADICAL OF AN L-IDEAL. If I is an ideal of any Banach algebra A, then Rad (I) (often called the kernel of the hull of I) is $\{a \in A : f(a) = 0$ for all $f \in \Delta(A)$ with $f(I) = 0\}$. For a closed ideal I, Rad (I) is just the set of elements $a \in A$ for which the image of a in A/I has spectral radius zero.

PROPOSITION. *If* $\mathfrak{N}$ *is an L-ideal of a commutative measure algebra* $\mathfrak{M}$, *then so is* Rad $(\mathfrak{N})$. *Also, with* J *the support of* $\mathfrak{N}$ *in* S *and* Rad $(J) = \{s \in S : f(s) = 0$ *for* $f \in \hat{S}$ *with* $f(J) = 0\}$, *we have that* Rad $(\mathfrak{N}) = \{\mu \in \mathfrak{M} : \mu$ *is concentrated on* Rad $(J)\}$.

PROOF. By definition, $\mu \in$ Rad $(\mathfrak{N})$ if and only if $\hat{\mu}(f) = 0$ for all $f \in \hat{S}$ with $f(J) = 0$. For such a measure, we also have $\hat{\mu}(fg) = \int fg \, d\mu = 0$ whenever $f(J) = 0$ and $g \in \hat{S}$. Since the span of $\hat{S}$ is dense in $C(S)$, we conclude that μ is concentrated on the ideal Rad (J) defined above. Hence, Rad $(\mathfrak{N}) = \{\mu \in \mathfrak{M}: \mu$ is concentrated on Rad $(J)\}$. This is clearly an L-subspace of $\mathfrak{M}$.

Note that it follows from the above that if Rad $(\mathfrak{N}) = \mathfrak{N}$ ($\mathfrak{N}$ is a primary ideal), then the support J of $\mathfrak{N}$ satisfies Rad $(J) = J$ and $\mathfrak{N}$ is determined by J as $\mathfrak{M} \cap M(J)$.

3.5.4. PRIME L-IDEALS. An L-ideal $\mathfrak{N} \subset \mathfrak{M}$ for which $\mathfrak{N}^\perp$ is a subalgebra (hence, an L-subalgebra) is called a prime L-ideal. Note that for such an ideal, the projection $\mathfrak{M} \to \mathfrak{N}^\perp$ is an algebra homomorphism. This projection, followed by the identity functional, yields a complex homomorphism $h \in \mathfrak{M}^*$. It follows that if we identify h with the corresponding element of $\hat{S}$, then $h^2 = h$ and

$$\mathfrak{N} = \{\mu \in \mathfrak{M} : \mu \text{ is concentrated on } h^{-1}(0)\},$$

$$\mathfrak{N}^\perp = \{\mu \in \mathfrak{M} : \mu \text{ is concentrated on } h^{-1}(1)\}.$$

A prime ideal $N \subset S$ is an ideal for which $S \backslash N$ is a subsemigroup. Note that if $h \in \hat{S}$ then $N_h = \{s \in S : h(s) = 0\}$ is such an ideal. In particular, if $h = h^2 \in \hat{S}$ is an idempotent, then N_h is an open-closed prime ideal with complement $S_h = \{s \in S : h(s) = 1\}$. Conversely, if N is an open-closed prime ideal, then $N = N_h$ and $S \backslash N = S_h$, where $h = h^2 \in \hat{S}$ is the characteristic function of $S \backslash N$.

Given an idempotent $h \in \hat{S}$, we may define an L-ideal $\mathfrak{N}_h$ and L-subalgebra $\mathfrak{M}_h$ of $\mathfrak{M}$ by

$$\mathfrak{N}_h = \{\mu \in \mathfrak{M} : \mu \text{ is concentrated on } N_h\},$$

$$\mathfrak{M}_h = \{\mu \in \mathfrak{M} : \mu \text{ is concentrated on } S_h\}.$$

Clearly, $\mathfrak{M} = \mathfrak{M}_h \oplus \mathfrak{N}_h$. Summarizing the above, we have

PROPOSITION. *There are one-to-one correspondences* $h \Longleftrightarrow N_h \Longleftrightarrow \mathfrak{N}_h$ *between idempotents* $h \in \hat{S}$, *open-closed prime ideals* $N_h \subset S$, *and prime L-ideals* $\mathfrak{N}_h \subset \mathfrak{M}$, *where*

(a) $N_h = \{s \in S : h(s) = 0\}$ *and* $S \backslash N_h = S_h = \{s \in S : h(s) = 1\}$; *and*

(b) $\mathfrak{N}_h = \{\mu \in \mathfrak{M} : \mu$ *is concentrated on* $N_h\}$ *and* $\mathfrak{N}_h^{\perp} = \mathfrak{M}_h = \{\mu \in \mathfrak{M} : \mu$ *is concentrated on* $S_h\}$.

By 3.5.2, the structure semigroup of $\mathfrak{N}_h$ is N_h. It is also worthwhile to identify the structure semigroup of the L-subalgebra $\mathfrak{M}_h$:

3.5.5. PROPOSITION. *For each idempotent* $h \in \hat{S}$, *the structure semigroup of* $\mathfrak{M}_h$ *is* S_h.

PROOF. By 3.3.3 and 3.3.4, it suffices to show that each complex homomorphism of $\mathfrak{M}_h$ extends to a complex homomorphism of $\mathfrak{M}$. However, this is trivial since for $f \in \Delta(\mathfrak{M}_h)$ the functional $\tilde{f} \in \Delta(\mathfrak{M})$ defined by $(\tilde{f}, \mu) = (f, \mu|_{S_h})$, provides such an extension.

CHAPTER 4. SPECIAL EXAMPLES

In this chapter we discuss the structure semigroups and spectra of several special measure algebras—notably $L(G), M(G)$, discrete semigroup algebras, and Arens-Singer algebras.

§4.1. Compactifications and structure semigroups

4.1.1. Suppose S is any abelian topological semigroup. Since $\hat{S} \cup \{0\}$ is closed under products and conjugation, the closure A of the linear span of $\hat{S}$ in supremum norm is a C^*-algebra of bounded continuous functions on S. It follows that the spectrum $\bar{S}$ of A is a compact Hausdorff space. Furthermore, since points of S determine complex homomorphisms of A, there is a continuous map $\alpha : S \to \bar{S}$, with dense image, such that $f \to f \circ \alpha : C(\bar{S}) \to A$ is an isomorphism-isometry of $C(\bar{S})$ onto A.

Note that each $f \in \hat{S}$ has the form $\tilde{f} \circ \alpha$ for a unique $\tilde{f} \in C(\bar{S})$. The elements $\tilde{f}$ for $f \in \hat{S}$ have dense linear span in $C(\bar{S})$; that is, they separate points in $\bar{S}$. If $\{s_\beta\}$ and $\{t_\beta\}$ are nets in S, with $\alpha(s_\beta) \to u$, $\alpha(t_\beta) \to v$ in $\bar{S}$, and if w is any cluster point of the net $\{\alpha(s_\beta t_\beta)\}$, then for $f \in \hat{S}$,

$$\tilde{f}(w) = \lim_\beta \tilde{f}(\alpha(s_\beta t_\beta)) = \lim_\beta f(s_\beta t_\beta) = \lim_\beta f(s_\beta) f(t_\beta) = \tilde{f}(u)\tilde{f}(v).$$

We draw three conclusions from this: (i) for $s, t \in S$ the element $\alpha(st)$ is uniquely determined by $\alpha(s)$ and $\alpha(t)$; (ii) if $\alpha(s_\beta) \to u$ and $\alpha(t_\beta) \to v$ for nets $\{s_\beta\}$, $\{t_\beta\}$ in S, then $\alpha(s_\beta t_\beta)$ converges to an element w depending only on u and v; (iii) if $f \in \hat{S}$ then its extension $\tilde{f}$ satisfies $\tilde{f}(w) = \tilde{f}(u)\tilde{f}(v)$ if u, v, and w are related as in (ii). This all adds up to the fact that $\bar{S}$ has a jointly continuous multiplication for which α is a homomorphism and each $\tilde{f}$, for $f \in \hat{S}$, is a semicharacter. Thus

4.1.2. Proposition. *If S is an abelian topological semigroup, then there is a unique compact topological semigroup $\bar{S}$ and continuous homomorphism $\alpha : S \to \bar{S}$, with dense image, such that*

(a) $\hat{\bar{S}}$ *separates points in* $\bar{S}$; *and*

(b) $f \to f \circ \alpha$ *is an isomorphism of* $\hat{\bar{S}}$ *onto* $\hat{S}$.

We shall call $\bar{S}$ (together with the map $\alpha : S \to \bar{S}$) the Bohr compactification of S. Note that α is injective if and only if $\hat{S}$ separates points in S.

For an l. c. a. group G, the Bohr compactification $\bar{G}$ is a compact group and, in fact, is the dual group of $(\hat{G}, d)$ – $\hat{G}$ with the discrete topology (Rudin [5]).

4.1.3. Proposition. *Let S be a locally compact abelian semigroup and* $\mathfrak{M} \subset M(S)$

a weak-∗ dense L-subalgebra. If each complex homomorphism of $\mathfrak{M}$ *has the form* $\mu \to \int f\,d\mu$ *for some* $f \in \hat{S}$, *then* $\bar{S}$ *is the structure semigroup of* $\mathfrak{M}$ *and the map* $\mu \to \mu_{\bar{S}}$ *of 3.2.3 is* $\mu \to \mu \circ \alpha^{-1}$, *where* $\alpha : S \to \bar{S}$ *is the canonical map of* S *into its Bohr compactification.*

PROOF. The semigroup $\bar{S}$ is compact; $\hat{\bar{S}}$ separates points, the image of $\mathfrak{M}$ in $M(\bar{S})$ under $\mu \to \mu \circ \alpha^{-1}$ is weak-∗ dense; and $\hat{\bar{S}}$ determines the spectrum of $\mathfrak{M}$. Since these properties characterize the structure semigroup of $\mathfrak{M}$, the proof is complete.

4.1.4. DISCRETE SEMIGROUP ALGEBRAS. The situation is particularly simple for measure algebras on discrete semigroups. An elementary conclusion is that if S is discrete, then each complex homomorphism of $M(S) = l^1(S)$ is given by an element of $\hat{S}$. In fact, if h is a complex homomorphism, then the function f defined by $f(s) = (h, \delta_s)$ is an element of $\hat{S}$ such that $(h, \mu) = \int f\,d\mu = \Sigma f(s)\mu(\{s\})$ for each $\mu \in M(S)$. We conclude from 4.1.3 that

COROLLARY. *If* S *is a discrete abelian semigroup, then the structure semigroup of* $M(S)$ *is the Bohr compactification of* S.

Although $M(S)$ is trivial from a measure theoretic point of view if S is discrete, it can be very interesting as a Banach algebra. Several very interesting function algebras can be described as the completion in spectral norm of a measure algebra on a discrete semigroup. In Chapter 10 we shall present examples of this sort which are important in the study of Gleason parts and the Shilov boundary.

§4.2. Group algebras

4.2.1. PROPOSITION. *If* G *is an l. c. a. group, then the structure semigroup of* $L(G)$ *is the Bohr compactification* $\bar{G}$ *of* G.

PROOF. Since the spectrum of $L(G)$ is $\hat{G}$, this follows immediately from 4.1.2.

4.2.2. THE RADICAL OF $L(G)$. Hewitt and Zuckerman [4] prove that if G is nondiscrete, then there is a positive, singular measure $\mu \in M(G)$ such that $\mu^2 \in L(G)$. Such a measure must clearly be an element of Rad $L(G)$ (3.5.3). Hence, $L(G)$ is a proper subideal of Rad $L(G)$ whenever G is nondiscrete.

Now if A is a commutative Banach algebra, B a subalgebra of A, and I an ideal with $I \subset B \subset \text{Rad}(I)$, then I and B have the same spectrum. In fact, since each element of $\text{Rad}(I)\backslash I$ has spectral radius zero, there are no complex homomorphisms of B which vanish on I. Thus, the restriction map $f \to f_I$ maps the spectrum of B homeomorphically onto the spectrum of I (3.5.1).

In the case where $A = M(G)$ and $I = L(G)$, we conclude

PROPOSITION. *If* $\mathfrak{N}$ *is any L-subalgebra of* $M(G)$ *such that* $L(G) \subset \mathfrak{N} \subset$ Rad $(L(G))$, *then* $\hat{G}$ *is the spectrum of* $\mathfrak{N}$ *and* $\bar{G}$ *is the structure semigroup of* $\mathfrak{N}$.

In Chapter 7 we shall prove a converse of the above result: every semisimple commutative measure algebra, for which the structure semigroup is a group, is an L-subalgebra of

$M(G)$, with $L(G) \subset \mathfrak{N} \subset \text{Rad } L(G)$, for some l.c.a. group G. The proof of this result is surprisingly difficult, requiring all of the machinery of Chapter 7.

4.2.3. *L*-IDEALS OF $L(G)$. There are no L-ideals of $L(G)$ other than (0) and $L(G)$. In fact, if $\mathfrak{N} \subset L(G)$ is an L-ideal, then $\mu \in \mathfrak{N}$ and $\gamma \in \hat{G}$ imply that $\gamma\mu \in \mathfrak{N}$. This implies that $\{\hat{\mu} : \mu \in \mathfrak{N}\}$ is a translation invariant algebra of functions on $\hat{G}$. It follows that if $\mathfrak{N} \neq (0)$ there is no $\gamma \in \hat{G}$ at which each $\hat{\mu}$, for $\mu \in \mathfrak{N}$, vanishes. That $\mathfrak{N} = L(G)$ in this case then follows from the Wiener-Tauberian theorem (Rudin [5]).

A more elementary proof that $L(G)$ contains no nonzero proper L-ideals is the following: if $\mathfrak{N} \neq 0$ is an L-ideal and $0 \neq \mu = fm \in \mathfrak{N}$, then $gm \in \mathfrak{N}$ for any $g \in L^1(G)$ which vanishes wherever f does. It follows that there is a bounded, compactly supported function k for which $km \in \mathfrak{N}$. However, $h = k * \tilde{k}$ is then a continuous function (Rudin [5]) with $hm \in \mathfrak{N}$. The fact that $\mathfrak{N}$ is an ideal of $L(G)$ implies it is translation invariant (since we can approximate point masses δ_g by elements of $L(G)$ in the weak-$*$ topology). Now for each compact set $K \subset G$, there is a finite sum of translates of $|h|$ which does not vanish on K. We conclude that for each compactly supported function in $L^1(G)$, the corresponding element of $L(G)$ is in $\mathfrak{N}$. Since $\mathfrak{N}$ is closed, $\mathfrak{N} = L(G)$.

In Chapter 7, we shall prove the converse: if $\mathfrak{M}$ is a semisimple commutative measure algebra with no proper, nonzero L-ideals, then $\mathfrak{M}$ is a group algebra.

§4.3. The algebra $M(G)$

4.3.1. There is not a great deal that can be said regarding the structure semigroup of $M(G)$, for nondiscrete G, other than that it is very complicated. Here we present a few of the elementary things that can be said.

Throughout this section G will be an l.c.a. group and S will be the structure semigroup of $M(G)$.

Since $L(G)$ is an L-ideal of $M(G)$ and has structure semigroup $\bar{G}$, the injection $L(G) \to M(G)$ induces an isomorphism of $\bar{G}$ onto an ideal of S (3.5.2). Since $\bar{G}$ is a group, it contains no proper ideals. Hence, the image of $\bar{G}$ in S must be the kernel of S. Since Rad $L(G)$ also has $\bar{G}$ as structure semigroup, it follows that Rad $L(G)$ consists of those measures $\mu \in M(G)$ for which μ_S is concentrated on the kernel of S (3.5.3). Thus

4.3.2. PROPOSITION. *The ideal* Rad $L(G)$ *consists of those measures* $\mu \in M(G)$ *for which* μ_S *is concentrated on the kernel of* S. *The kernel of* S *may be identified with* $\bar{G}$ *in such a way that the map* $\mu \to \mu_S$, *for* $\mu \in \text{Rad } L(G)$, *is just the map of measures induced by the natural map of* G *into its Bohr compactification* $\bar{G}$.

4.3.3. THE IMAGE OF $\hat{G}$ IN $\hat{S}$. The elements of the spectrum of $M(G)$ determined by characters in $\hat{G}$ are exactly those which do not vanish on $L(G)$ (3.5.1). It is apparent from 4.3.2 that $f \in \hat{S}$ determines a complex homomorphism of $M(G)$ vanishing on $L(G)$ if and only if f is zero on the kernel K of S. However, the restriction of f to K is either zero or is a character of K, and the latter happens if and only if $|f| \equiv 1$ (3.4.6). Hence

COROLLARY. *The canonical image of* $\hat{G}$ *in* $\hat{S}$ *is* $\{f \in \hat{S} : |f| = 1\}$, *which is the maximal group at the identity in* $\hat{S}$.

Note that if multiplication were jointly continuous in $\hat{S}$ (in the Gelfand topology) then $\hat{G}$ would be closed and, hence, compact (3.4.7). Similarly, if the map $f \to |f|$ were continuous, then $\hat{G}$ would be compact. Since $\hat{G}$ is compact if and only if G is discrete, we conclude that $(f, g) \to fg$ and $f \to |f|$ are not continuous operations in $\hat{S}$ for G nondiscrete.

4.3.4. SOME PRIME *L*-IDEALS OF $M(G)$. Let $\mathcal{E}$ be a family of Borel subsets of G which is closed under translation, $E \to gE$, and multiplication, $(E, F) \to E \cdot F$. It is easy to see that if $\mathfrak{N} = \{\mu \in M(G) : |\mu|(E) = 0$ for all $E \in \mathcal{E}\}$, then $\mathfrak{N}$ is an L-subspace with $\mathfrak{N}^\perp = \{\nu \in M(G) : \nu$ is concentrated on a countable union of sets in $\mathcal{E}\}$. Furthermore, $\mathfrak{N}$ is an ideal, since for $E \in \mathcal{E}, \mu \in \mathfrak{N}$, and $\nu \in M(G)$ we have

$$\nu * \mu(E) = \int \mu(g^{-1}E)\, d\nu(g) = 0.$$

The fact that $\mathcal{E}$ is closed under products implies that $\mathfrak{N}^\perp$ is a subalgebra of $M(G)$. We conclude that $\mathfrak{N}$ is a prime L-ideal of $M(G)$ (3.5.4); thus, $\mathfrak{N} = \{\mu \in M(G) : \mu_S$ is concentrated on $N_h\}$ for some idempotent $h \in \hat{S}$, while $\mathfrak{N}^\perp = \{\mu \in M(G) : \mu_S$ is concentrated on $S_h\}$.

4.3.5. TOPOLOGIES ON G. Suppose G_τ is an l. c. a. group continuously isomorphic to G; that is, G_τ is G as a group, but is an l. c. a. group under a topology τ possibly stronger than that of G. Then the compact sets of G_τ form a subclass of the class of compact sets of G. The algebra $M(G_\tau)$ may be identified as that subalgebra of $M(G)$ consisting of measures concentrated on a countable union of compact sets of G_τ.

If $\mathfrak{N}_\tau = \{\mu \in M(G) : |\mu|(K) = 0$ for each compact set K of $G_\tau\}$, then $\mathfrak{N}_\tau^\perp = M(G_\tau)$. Since the compact sets of G_τ are closed under translation and products, we conclude from 4.3.4 that

PROPOSITION. *If* G_τ *is an l. c. a. group continuously isomorphic to* G, *then there is an idempotent* $h \in \hat{S}$ *such that the structure semigroup of* $M(G_\tau)$ *is* S_h *and* $M(G_\tau) = \{\mu \in M(G) : \mu_S$ *is concentrated on* $S_h\}$.

Note that by 4.3.2 the kernel K_h of the subsemigroup S_h is isomorphic to $\overline{G}_\tau$ and Rad $L(G_\tau)$ is the set of $\mu \in M(G)$ such that μ_S is concentrated on K_h.

4.3.6. THE DISCRETE TOPOLOGY. In the case of the discrete topology d on G, we have that $M(G_d) = L(G_d)$. Thus, the subsemigroup S_h, as above, must be the group $\overline{G}_d$. Since the identity δ of $M(G)$ maps to δ_e under $\mu \to \mu_S$ (where e is the identity of S) and since $\delta \in M(G_d)$, we conclude that S_h is a group containing the identity of S. Since its complement N_h is an ideal, S_h must be the maximal group at the identity in S. Thus

PROPOSITION. *The maximal group at the identity in* S *is isomorphic to* $\overline{G}_d$. *Its complement is an open-closed prime ideal. Also* $M(G_d)$ *consists of those measures* $\mu \in M(G)$ *such that* μ_S *is concentrated on this group.*

4.3.7. INDEPENDENT SETS IN S. According to 1.3.4, if G is nondiscrete there is a Cantor set $E \subset G$ such that each element of the unit ball of $M_c(E)^*$ extends to a complex homomorphism of $M(G)$. Let $\mathfrak{N}$ be the image of $M_c(E)$ in $M(S)$ under $\mu \to \mu_S$ and let $D \subset S$ be the smallest closed set on which each element of $\mathfrak{N}$ is concentrated. Then every element of the unit ball of $\mathfrak{N}^*$ is determined by the restriction to D of an element of $\hat{S}$.

If we recall the properties which characterize the standard representation of an L-space (2.4.2), we conclude that $\mu \to \mu_S : M_c(E) \to M(D)$ must be the standard representation of $M_c(E)$. Hence

PROPOSITION. *If G is nondiscrete, then $M(G)$ contains a nonzero L-subspace consisting of continuous measures, such that the support D of its image in $M(S)$ is extremely disconnected and has the property that every function $f \in C(D)$, with $|f| \leqslant 1$, is the restriction of an element of $\hat{S}$.*

This has the following consequence.

COROLLARY. *If G is nondiscrete, S contains a semicharacter f for which $|f|$ has values other than zero and one. Hence S is not a union of groups.*

§4.4. Arens-Singer algebras

4.4.1. Let G be an l.c.a. group and T a closed subsemigroup of G which has dense interior and generates G. Then $L(T) = L(G) \cap M(T)$ is a weak-$*$ dense L-subalgebra of $M(T)$.

Algebras of the form $L(T)$ were studied by Arens and Singer [1]. We describe some of their conclusions below.

PROPOSITION. *The spectrum of $L(T)$ is $\hat{T}$ and the structure semigroup is $\overline{T}$.*

PROOF. The dual space of $L(T)$ is $L^\infty(m|_T)$. Furthermore, an argument like that in 1.1.6 shows that $f \in L^\infty(m|_T)$ determines a complex homomorphism of $L(T)$ if and only if f is equal almost everywhere to a function which is continuous and multiplicative. It follows from 4.1.3 that the structure semigroup of $L(T)$ is the Bohr compactification of T.

4.4.2. THE SHILOV BOUNDARY OF $L(T)$. If B is a closed subalgebra of a commutative Banach algebra A, then every complex homomorphism in the Shilov boundary of B extends to a complex homomorphism of A (Gamelin [1]). If we apply this to the subalgebra $L(T)$ of $L(G)$, we conclude that every element of $\hat{T}$ which is in the Shilov boundary of $L(T)$ is the restriction to T of an element of $\hat{G}$.

If $f \in \hat{T}$ and $|f| = 1$, then the map $\mu \to f\mu : L(T) \to L(T)$ is an automorphism with inverse $\mu \to \bar{f}\mu$. The homeomorphism of the spectrum of $L(T)$ induced by this automorphism is just $g \to fg : \hat{T} \to \hat{T}$. Clearly the Shilov boundary must be left invariant by this map. Hence, if g is in the Shilov boundary for $L(T)$, then so is fg for every $f \in \hat{T}$ with $|f| = 1$.

We now have that the Shilov boundary for $L(T)$ is a subset of $\{\gamma|_T : \gamma \in \hat{G}\} \subset \{f \in \hat{T} : |f| = 1\}$ which is invariant under multiplication by $\{f \in \hat{T} : |f| = 1\}$. It follows that:

PROPOSITION. *The following three sets are identical: the Shilov boundary of* $L(T)$, $\{\gamma|_T : \gamma \in \hat{G}\}$, *and* $\{f \in \hat{T} : |f| = 1\}$.

Arens and Singer [1] also discuss the polar decomposition $f = h|f|$ (3.4.4) and the analytic structure in $\hat{T}$ induced by the maps $z \to h|f|^z$ (Re $(z) > 0$). They also discuss representing measures on $\hat{G}$ for points of $\hat{T}$.

4.4.3. EXAMPLES. A simple example of an algebra of the Arens-Singer type is obtained by letting $G = Z$ and $T = Z^+ = \{n \in Z : n \geqslant 0\}$. Here, $\hat{T}$ may be identified with the unit disc $D \subset \mathbf{C}$, where $z \in D$ determines the semicharacter $n \to z^n$. The Gelfand transform associates to $\mu \in M(Z^+)$ the analytic function

$$\hat{\mu}(z) = \sum_{n=0}^{\infty} \mu(\{n\})z^n.$$

Another example is obtained by letting $G = R$ and $T = R^+$. Here $\hat{T}$ is the right half plane in $\mathbf{C}$ and the Gelfand transform of $\mu \in L(R^+)$ is the Laplace transform

$$\hat{\mu}(z) = \int e^{-zt}\, d\mu(t).$$

A less trivial example is obtained by letting G be the multiplicative group of positive rationals and letting T be the subsemigroup consisting of positive integers. The algebra $L(T)$ is related to Dirichlet series (cf. Hewitt and Williamson [1]).

In Chapter 10 we shall discuss another example—one which shows that one-point Gleason parts need not be in the Shilov boundary.

§4.5. An existence theorem for semicharacters

4.5.1. For lack of a better place, we present here a theorem on existence of semicharacters on a closed subsemigroup T of an l.c.a. group G. This theorem implies, in the case where T has dense interior, that for an Arens-Singer algebra on $T \subset G$, there are complex homomorphisms other than those determined by points of $\hat{G}$. In this form, the theorem is due to Rieffel [1]; however, the version presented here is somewhat more general. The proof involves Banach algebra theory, although the statement of the theorem does not.

4.5.2. THEOREM. *Let* T *be a closed subsemigroup of an l.c.a. group* G *such that* T *is not contained in any proper closed subgroup of* G. *Then* 1 *is a limit point of* $\{f \in \hat{T} : 0 < f < 1\}$ *in the compact-open topology.*

We prove this using the structure theory for l.c.a. groups (Rudin [5]). The proof is broken into several lemmas.

4.5.3. LEMMA. *The conclusion of* 4.5.2 *holds if* G *is discrete.*

PROOF. Suppose 1 is isolated in $\hat{T}^+ = \{f \in \hat{T} : f \geqslant 0\}$. Then $\{f \in \hat{T} : |f| = 1\}$ is both open and compact in $\hat{T}$. Since $\hat{T}$ is the spectrum of $l^1(T) = M(T)$ (4.1.4), it follows from the Shilov idempotent theorem 1.4.2 that there is an idempotent $\nu \in M(T)$ such that $\hat{\nu}(f) = 1$ if $|f| = 1$ and $\hat{\nu}(f) = 0$ if $|f| < 1$ for each $f \in \hat{T}$. It follows that $\nu = \delta_e$, the point mass at the identity of G. Hence, $e \in T$.

Now T generates G but does not equal G. Hence, for some $g \in T$, $g^{-1} \notin T$ and thus $e \notin gT$. It follows from the above paragraph, applied to gT, that $1 = \lim_\alpha f_\alpha$ for a net $\{f_\alpha\} \subset (gT)^\wedge$ with $0 \leqslant f_\alpha < 1$. However, for each α such that $f_\alpha(g) \neq 0$, $\widetilde{f}_\alpha(g_1) = f_\alpha(g_1 g) f_\alpha^{-1}(g)$ defines a positive semicharacter on T. The lemma follows.

4.5.4. LEMMA. *Let $T \subset G$ be as in 4.5.2. If G has a compact subgroup G_0 such that $G/G_0 \simeq R^n$, then there is a homomorphism $\beta: G \to R$ such that $\beta(T) \subset R^+$ and $\beta(T) \neq (0)$.*

PROOF. Let $g \in T$ be such that $g^{-1} \notin T$. We claim that $g^{-1} \notin G_0 T$. In fact, if $g^{-1} = g_0 g_1$ with $g_0 \in G_0$ and $g_1 \in T$, then $g_0^{-1} = g g_1 \in T \cap G_0$. However, $G_0 \cap T$ is a closed subsemigroup of the compact group G_0 and, hence, must be a subgroup. This forces g_0 and g^{-1} to be elements of T, which is a contradiction.

Now $G_0 T$ is a closed subsemigroup of G which is not G but generates G. It follows that its image $\widetilde{T}$ in $G/G_0 \simeq R^n$ has the same properties as a subsemigroup of R^n. We complete the proof by verifying the lemma for the semigroup $\widetilde{T} \subset R^n$.

Now if $u, v \in \widetilde{T} \cup \{0\}$, $w \in \widetilde{T}$, but $-w \notin \widetilde{T}$, then we claim that the line segment $\{tu + (1-t)v : 0 \leqslant t \leqslant 1\}$ cannot contain $-w$. In fact, if $-w = tu + (1-t)v$, then by approximating T by rational numbers p/q, $0 \leqslant p \leqslant q$, we conclude that $-w$ is a limit of elements of $\widetilde{T} \cup \{0\}$ of the form $(q-1)w + pu + (q-p)v$. Since $\widetilde{T}$ is closed, this cannot happen. We conclude that the convex hull of $T \cup \{0\}$ is a cone which does not contain $-w$. It follows that there is a linear functional on R^n which is nonnegative on $\widetilde{T}$ and negative at $-w$ (hence, positive at w). This functional, composed with $G \to G/G_0 \simeq R^n$, is the homomorphism β that we seek.

4.5.5. LEMMA. *Let T be a closed subsemigroup of the l. c. a. group G containing the identity, and H an open subgroup of G. Then each element of $(H \cap T)^\wedge$ extends to an element of $\hat{T}$.*

PROOF. Consider the algebra $M_d(T) = l^1(T)$ and its subalgebra $M_d(T \cap H)$. It follows from the next lemma that every complex homomorphism of the second algebra extends to a complex homomorphism of the first.

Now if $f \in (T \cap H)^\wedge$ then f determines a complex homomorphism of $M_d(T \cap H)$. An extension of this to a complex homomorphism of $M_d(T)$ will be determined by a bounded semicharacter $\widetilde{f}$ on T with $\widetilde{f}|_{T \cap H} = f$. It follows that $\widetilde{f}$ is continuous on the intersection with T of each coset of H. Since each of these sets is open in T, we conclude that $\widetilde{f}$ is continuous and $\widetilde{f} \in \hat{T}$.

4.5.6. LEMMA. *Let $\mathfrak{M}$ be a measure algebra with identity and $\mathfrak{N} \subset \mathfrak{M}$ an L-subalgebra containing the identity. If $\mathfrak{N} * \mathfrak{N}^\perp \subset \mathfrak{N}^\perp$ then every complex homomorphism of $\mathfrak{N}$ extends to a complex homomorphism of $\mathfrak{M}$.*

PROOF. To prove this, it suffices to show that if $\mu_1, \cdots, \mu_n \in \mathfrak{N}$, then the joint spectrum of the tuple $(\mu_1, \cdots, \mu_n)$ in $\mathfrak{M}$ is the same as its joint spectrum in $\mathfrak{N}$. This is equivalent to proving that each such tuple which is nonsingular in $\mathfrak{M}$ is also nonsingular in $\mathfrak{N}$.

Thus, let $\mu_1, \cdots, \mu_n \in \mathfrak{N}$ be such that the equation

(i) $$\mu_1 * \nu_1 + \cdots + \mu_n * \nu_n = \delta$$

has a solution $\nu_1, \cdots, \nu_n \in \mathfrak{M}$. If we decompose each ν_i as $\omega_i + \rho_i$ with $\omega_i \in \mathfrak{N}$, $\rho_i \in \mathfrak{N}^\perp$, then the condition $\mathfrak{N} * \mathfrak{N}^\perp \subset \mathfrak{N}^\perp$ implies that $\mu_1 * \rho_1 + \cdots + \mu_n * \rho_n = 0$ and

$$\mu_1 * \omega_1 + \cdots + \mu_n * \omega_n = \delta.$$

Hence (i) also has a solution in $\mathfrak{N}$. This completes the proof.

4.5.7. PROOF OF 4.5.2. By the structure theory for l. c. a. groups (Rudin [5]), G has an open subgroup H such that H is the product of R^n $(n \geqslant 0)$ with a compact group.

Now if T does not meet each coset of H, then its image in the discrete group G/H is a proper subsemigroup which generates G/H. The proposition then follows directly from Lemma 4.5.3 applied to G/H.

If T does meet each coset of H, then $T \cap H$ must be a proper closed subsemigroup of H which generates H. If $\beta : H \to R$ is the map of Lemma 4.5.4, then $\{e^{-t\beta}; t > 0\}$ is a family of strictly positive elements of $(T \cap H)^\wedge$ with 1 as a limit point. For each t let $f_t \in \hat{T}$ be an extension of $e^{-t\beta}$ as in Lemma 4.5.5. If $g \in T$, then $hg^{-1} = g_1$ for some $h \in H$, $g_1 \in T$ (since T meets each coset of H). Then $h \in T \cap H$ and $e^{-t\beta(h)} = f_t(h) = f_t(g) f_t(g_1)$. Since $|f_t| \leqslant 1$, it follows that $\lim_{t \to 0} f_t(g) = 1$. This convergence is uniform on compact subsets of H and, hence, on compact subsets of cosets of H. Since cosets of H are open, the convergence is uniform on all compact subsets of G. This completes the proof.

4.5.8. COROLLARY. *For an Arens-Singer algebra* $L(T)$, *the Shilov boundary* $\hat{G}|_T$ *of* $L(T)$ *is not the entire spectrum* $\hat{T}$; *in fact,* $\hat{T}$ *contains nonnegative semicharacters distinct from* 1 *in every neighborhood of* 1.

§4.6. Idempotent semigroup algebras

4.6.1. An idempotent semigroup is one in which each element is idempotent. An abelian idempotent semigroup can also be described as a semilattice, that is, an ordered set in which each pair of elements s, t has a least upper bound $s \vee t$. The semigroup operation is $(s, t) \to s \vee t$.

Many interesting measure algebras arise as L-subalgebras of the measure algebra on an idempotent semigroup.

4.6.2. $M[0, 1]$ AND $L[0, 1]$. If the interval $[0, 1]$ is given its usual topology and the multiplication $(x, y) \to x \vee y$, then it becomes a compact idempotent semigroup. The corresponding measure algebra $M[0, 1]$ has been studied by Hewitt and Zuckerman [3].

Each complex homomorphism of $M[0, 1]$ is determined by a function on $[0, 1]$ of the form $\chi_{[0,t]}$ or $\chi_{[0,t)}$ for $t \in [0, 1]$. These functions are exactly the (discontinuous) semicharacters on $[0, 1]$. Thus, the spectrum of $M[0, 1]$ can be described as follows: consider the set Δ consisting of 0 and the set of symbols of the form $t+$ and $t-$ for $t \in (0, 1]$. This becomes a totally ordered set if we declare that $0 < s\pm < t\pm$ if $0 < s < t$ and that $t- < t+$. Under the order topology Δ is compact. We may identify it with the spectrum of $M[0, 1]$ if we define the Gelfand transform of $\mu \in M[0, 1]$ by

$$\hat{\mu}(0) = \mu(\{0\}),$$

$$\hat{\mu}(t-) = \int \chi_{[0,t)}\, d\mu = \mu([0, t)),$$

and

$$\hat{\mu}(t+) = \int \chi_{[0,t]}\, d\mu = \mu([0, t]).$$

Note that $\hat{\mu}(t-) = \hat{\mu}(t+)$ for all but a countable set of t and that $\hat{\mu}$ is determined by its values on $\{0\} \cup \{t+ : t \in (0, 1]\}$. When restricted to this set, $\hat{\mu}$ may be considered a function on $[0, 1]$. In fact, $\mu \to \hat{\mu}$ determines an isomorphism-isometry of $M[0, 1]$ onto the algebra of functions of bounded variation on $[0, 1]$ with pointwise multiplication and total variation norm.

The structure semigroup for $M[0, 1]$ has the following description. Let S be the set consisting of symbols $0, 0+, 1-, 1$, and $t-, t, t+$ for $t \in (0, 1)$. We give S the obvious total order and the order topology. Then S is a compact idempotent semigroup under max multiplication. To the discontinuous semicharacters $\chi_{[0,t)}$ and $\chi_{[0,t]}$ on $[0, 1]$ we associate the continuous semicharacters $\chi_{[0,t-]} = \chi_{[0,t)}$ and $\chi_{[0,t]} = \chi_{[0,t+)}$ on S. This identifies the spectrum of $M[0, 1]$ with $\hat{S}$.

The spaces $M_c[0, 1]$ and $L[0, 1]$ are L-subalgebras of $M[0, 1]$. For each of these algebras the spectrum is $(0, 1]$. That is, complex homomorphisms are determined by the discontinuous semicharacter $\chi_{[0,t]}$ for $t > 0$ (note that $\chi_{[0,t)} = \chi_{[0,t]}$ a.e. with respect to any continuous measure). The Gelfand transform maps $M_c[0, 1]$ onto the space of continuous functions of bounded variation and $L[0, 1]$ onto the space of absolutely continuous functions. In either case, the structure semigroup is as before except that the unsigned member of each triple $\{t-, t, t+\}$ is deleted.

4.6.3. Generalizations. An analysis similar to the one above can be carried out for measure algebras on any totally ordered space. Such algebras were studied by Ross [1].

Note that for $M[0, 1]$ the structure semigroup is again an idempotent semigroup. This also turns out to be the case for measure algebras on finite products of $[0, 1]$ with itself (Newman [1]). However, it is not true in general that if T is an idempotent semigroup then $M(T)$ has a structure semigroup which is idempotent. In fact, if T is an infinite product of copies of $[0, 1]$ or $\{0, 1\}$, then $M(T)$ contains measures with mutually singular powers (Newman [1]). It follows that the structure semigroup of $M(T)$ cannot be idempotent.

Newman [1] gives several conditions which guarantee that the structure semigroup of a measure algebra $\mathfrak{M}$ is idempotent. One such condition is that the spectrum of each nonnegative measure in $\mathfrak{M}$ be nonnegative.

In another paper on the subject, Newman [2] generalizes the connection between $M[0, 1]$ and functions of bounded variation. On any idempotent semigroup he defines the notion of a function of bounded variation and defines a bounded variation norm. The space of functions of bounded variation turns out to be a Banach algebra under pointwise multiplication and, in fact, a convolution measure algebra.

CHAPTER 5. THE STRUCTURE OF $\hat{S}$

Throughout this section $\mathfrak{M}$ will be a semisimple commutative measure algebra with an identity of norm one. We identify $\mathfrak{M}$ with its image $\mathfrak{M}_S \subset M(S)$, where S is the structure semigroup of $\mathfrak{M}$

The spectrum $\hat{S}$ of $\mathfrak{M}$ has rather special algebraic and topological properties. This chapter is devoted to studying some of these properties.

§5.1. **Topologies on $\hat{S}$**

5.1.1. Since we have assumed that $\mathfrak{M}$ has an identity of norm one, S has an identity e (3.4.2) and $\hat{S}$ is closed under pointwise products. Furthermore, $\hat{S}$ is compact in the Gelfand topology.

We have already noted that the multiplication $(f, g) \to fg$ in $\hat{S}$ is separately continuous but not, in general, jointly continuous for the Gelfand topology induced on $\hat{S}$ by $\mathfrak{M}$. It turns out to be very useful to introduce a new topology on $\hat{S}$ for which multiplication is jointly continuous. Before defining this topology, we make the following observation.

5.1.2. PROPOSITION. *Under the operation* $(f, \mu) \to f\mu : \hat{S} \times \mathfrak{M} \to \mathfrak{M}$, $\hat{S}$ *acts as a semigroup of bounded endomorphisms of* $\mathfrak{M}$.

PROOF. This operation certainly represents $\hat{S}$ as a semigroup of bounded linear operators on $\mathfrak{M}$. Hence, we need only prove that $f(\mu * \nu) = (f\mu) * (f\nu)$ for $\mu, \nu \in \mathfrak{M}$, $f \in \hat{S}$. However, for $g \in C(S)$,

$$\int g\, d(f(\mu * \nu)) = \int gf\, d\mu * \nu = \iint g(st) f(st)\, d\mu(s)\, d\nu(t)$$
$$= \iint g(st) f(s) f(t)\, d\mu(s)\, d\nu(t) = \int g\, d(f\mu * f\nu).$$

Thus $f(\mu * \nu) = (f\mu) * (f\nu)$.

5.1.3. THE STRONG TOPOLOGY. With the above representation of $\hat{S}$ as a semigroup of endomorphisms of $\mathfrak{M}$, the Gelfand topology for $\hat{S}$ can be described as the weak operator topology, that is, the topology in which $f_\alpha \to f$ if and only if $(g, f_\alpha\mu) \to (g, f\mu)$ for each $\mu \in \mathfrak{M}$ and $g \in \mathfrak{M}^*$. In fact, the equality $(g, f\mu) = (f, g\mu) = (g\mu)\,\hat{}\,(f)$, for $f \in \hat{S}$, $g \in \mathfrak{M}^*$, $\mu \in \mathfrak{M}$, implies that the weak operator and Gelfand topologies agree.

Another natural topology on a family of operators is the strong operator topology. This topology, for $\hat{S}$ acting on $\mathfrak{M}$, is the topology in which $f_\alpha \to f$ if and only if $f_\alpha\mu \to f\mu$ in norm for each $\mu \in \mathfrak{M}$.

TERMINOLOGY. The strong operator topology for $\hat{S}$ acting on $\mathfrak{M}$ will, henceforth, be called the strong topology of $\hat{S}$. Similarly, the Gelfand or weak operator topology of $\hat{S}$ will, henceforth, be called the weak topology. When convenient, we shall denote $\hat{S}$ with the strong topology by $\hat{S}_\sigma$ and $\hat{S}$ with the weak topology by $\hat{S}_\omega$.

By definition, a typical neighborhood of $f \in \hat{S}$ in the strong topology has the form $\{g \in \hat{S} : \|f\mu_i - g\mu_i\| < \epsilon, i = 1, \cdots, n\}$ for measures $\mu_1, \cdots, \mu_n \in \mathfrak{M}$ and $\epsilon > 0$. If we set $\mu = |\mu_1| + \cdots + |\mu_n|$, then this neighborhood contains the neighborhood $\{g \in \hat{S} : \|f\mu - g\mu\| = \int |f - g| \, d\mu < \epsilon\}$. Hence, the sets of this latter form, for $0 \leqslant \mu \in \mathfrak{M}$ and $\epsilon > 0$, form a base for the strong topology.

5.1.4. PROPOSITION. *The following operations are continuous*:

(a) $(f, g) \to fg : \hat{S}_\sigma \times \hat{S}_\sigma \to \hat{S}_\sigma$;

(b) $(f, g) \to fg : \hat{S}_\sigma \times \hat{S}_\omega \to \hat{S}_\omega$;

(c) $f \to \bar{f} : \hat{S}_\sigma \to \hat{S}_\sigma$ *and* $f \to \bar{f} : \hat{S}_\omega \to \hat{S}_\omega$;

(d) $f \to |f| : \hat{S}_\sigma \to \hat{S}_\sigma$.

PROOF. Part (a) follows from the inequality

$$\begin{aligned}\int |fg - f_1 g_1| \, d\mu &\leqslant \int |f| \, |g - g_1| \, d\mu + \int |g_1| \, |f - f_1| \, d\mu \\ &\leqslant \int |g - g_1| \, d\mu + \int |f - f_1| \, d\mu\end{aligned}$$

for $f, f_1, g, g_1 \in \hat{S}$ and $0 \leqslant \mu \in \mathfrak{M}$. Similarly, part (b) follows from the fact that for $f, f_1, g, g_1 \in \hat{S}$ and $\mu \in \mathfrak{M}$,

$$\begin{aligned}|\hat{\mu}(fg) - \hat{\mu}(f_1 g_1)| &\leqslant |\hat{\mu}(fg) - \hat{\mu}(fg_1)| + |\hat{\mu}(fg_1) - \hat{\mu}(f_1 g_1)| \\ &\leqslant |(f\mu)\,\hat{}\,(g) - (f\mu)\,\hat{}\,(g_1)| + \|f\mu - f_1\mu\|,\end{aligned}$$

which implies that the product of the strong neighborhood $\{f_1 : \|f\mu - f_1\mu\| < \epsilon/2\}$ and the weak neighborhood $\{g_1 : |(f\mu)\,\hat{}\,(g) - (f\mu)\,\hat{}\,(g_1)| < \epsilon/2\}$ is contained in the weak neighborhood $\{h : |\hat{\mu}(fg) - \hat{\mu}(h)| < \epsilon\}$.

Part (c) is obvious, since $\hat{\mu}(\bar{f}) = (\bar{\mu}\,\hat{}\,(f))^-$ and $\|\bar{f}\mu\| = \|f\mu\|$. Part (d) follows from the inequality

$$\int \big| |f| - |g| \big| \, d\mu \leqslant \int |f - g| \, d\mu$$

for $\mu \geqslant 0$.

5.1.5. COMPARISON OF S_σ AND S_ω. Obviously, the strong topology dominates the weak topology on $\hat{S}$. The weak topology has the advantage that $\hat{S}_\omega$ is compact, while the strong topology has the advantage that the operations on $\hat{S}_\sigma$ are all continuous. Fortunately for our later arguments, the two topologies agree on certain kinds of subsets of $\hat{S}$.

We set $\hat{S}^+ = \{f \in \hat{S} : f \geqslant 0\}$.

PROPOSITION. (a) *If* $\{f_\alpha\}$ *is a net in* $\hat{S}$, $f \in \hat{S}$, *and* $|f_\alpha| \leqslant |f|$ *for every* α, *then* $f_\alpha \to f$ *strongly if and only if* $f_\alpha \to f$ *weakly*;

(b) *the weak and strong topologies agree on any set of the form* $\{f \in \hat{S} : |f| = g\}$ $(g \in \hat{S}^+)$;

(c) *the weak and strong topologies agree on any subset of* $\hat{S}^+$ *which is totally ordered.*

PROOF. Part (a) follows from the sequence of inequalities (for $0 \leqslant \mu \in \mathfrak{M}$ and $|f_\alpha| \leqslant |f|$):

$$\|f\mu - f_\alpha \mu\|^2 = [\int |f - f_\alpha|\, d\mu]^2 \leqslant \|\mu\| \int |f - f_\alpha|^2\, d\mu$$
$$\leqslant \|\mu\| \int (2|f|^2 - f\bar{f}_\alpha - \bar{f} f_\alpha)\, d\mu = 2\,\|\mu\|\, \mathrm{Re}\, [(\bar{f}\mu)^\wedge (f) - (\bar{f}\mu)^\wedge (f_\alpha)] .$$

Part (b) follows immediately from (a). Part (c) is a consequence of the fact that for $g, f \in \hat{S}^+, g \leqslant f,$ and $0 \leqslant \mu \in \mathfrak{M}$,

$$\|f\mu - g\mu\| = \int |f - g|\, d\mu = \int (f - g)\, d\mu = \hat{\mu}(f) - \hat{\mu}(g).$$

5.1.6. CLOSED SETS IN $\hat{S}^+$. Note that $\hat{S}^+$ is closed in both the strong and weak topologies, as is each of its subsets of the form $\{f \in \hat{S}^+ : g \leqslant f \leqslant h\}$. This is due to the fact that f and g are in $\hat{S}^+$ and $f \leqslant g$ if and only if $0 \leqslant \hat{\mu}(f) \leqslant \hat{\mu}(g)$ for every positive $\mu \in \mathfrak{M}$.

PROPOSITION. *If* $\Omega \subset \hat{S}^+$ *is closed in the strong topology, then* Ω *contains minimal and maximal elements.*

PROOF. Let $F \subset \Omega$ be any totally ordered subset. Since $\hat{S}^+$ is weakly compact, there is an element $g \in S^+$ which is a weak limit point of F and satisfies $g \leqslant f$ for every $f \in F$. By 5.1.5(c) the weak and strong topologies agree on the totally ordered set $F \cup \{g\}$. It follows that g is also a strong limit point of F and, hence, an element of F. That Ω has a minimal element now follows from Zorn's lemma.

A similar argument shows that Ω has a maximal element.

It turns out that if Ω is both open and closed in the strong topology, then a minimal element of Ω must have a very special form (5.2.4).

§5.2. Ideals, groups, and critical points

5.2.1. IDEALS. Since $\hat{S}$ is a semigroup, we may consider its ideals, idempotents, and maximal groups (§3.4). The ideals have an interesting characterization in terms of the order relation

5.2.2. PROPOSITION. (a) *For* $g \in \hat{S}$, *the principal ideal* $g\hat{S}$ *is just* $\{f \in \hat{S} : |f| \leqslant |g|\}$;

(b) *If* $A \subset \hat{S}$ *then* A *is an ideal if and only if* $g \in A, f \in \hat{S},$ *and* $|f| \leqslant |g|$ *imply* $f \in A$.

PROOF. Clearly $g\hat{S} \subset \{f \in \hat{S} : |f| \leqslant |g|\}$. The reverse containment follows from 3.4.4. In fact, if $|f| \leqslant |g|$, then the function h_1, defined by

$$h_1(s) = f(s)g^{-1}(s) \qquad (g(s) \neq 0),$$
$$h_1(s) = 0 \qquad (g(s) = 0),$$

is bounded, Borel, and multiplicative. By 3.4.4 it is equal a.e. to an element $h \in \hat{S}$ which obviously satisfies $f = hg$.

Part (b) follows immediately from (a).

Note that a principal ideal $g\hat{S}$ is necessarily weakly closed (hence, strongly closed), since it is the image of the compact set $\hat{S}$ under the continuous map $f \to gf$.

5.2.3. IDEMPOTENTS AND MAXIMAL GROUPS. If $h = h^2 \in \hat{S}$ is idempotent, then the maximal group in $\hat{S}$ containing h is $\Gamma_h = \{f \in \hat{S} : |f| = h\}$. Conjugation is inversion in this group.

Note that Γ_h is strongly closed since $\hat{S}_\sigma$ is a topological semigroup (3.4.7), but it is not generally weakly closed. However, when restricted to Γ_h, the weak and strong topologies agree by 5.1.5(b). Hence, Γ_h is a topological group in the weak (strong) topology. Note that Γ_h is locally compact if and only if it is (weakly) open in its (weak) closure in $\hat{S}$. This happens, in particular, if h is a critical point as defined in 5.2.4.

Note that h is an identity (and Γ_h the maximal group at the identity) for the closed ideal $h\hat{S}$.

Each element $h \in \hat{S}^+$ is on a totally ordered arc with endpoints which are idempotents. In fact, the strong closure of the set $\{h^r : 0 < r < \infty\}$ contains the idempotents $h^0 = \lim_{r \to 0} h^r$ and $h^\infty = \lim_{r \to \infty} h^r$. Note that h^0 is the element of $\hat{S}^+$ which is equal a.e. $/\mathfrak{M}$ to the Borel semicharacter k^0 defined by $k^0(s) = 1$ if $h(s) > 0$ and $k^0(s) = 0$ if $h(s) = 0$ (3.4.4). Similarly, h^∞ is the element of $\hat{S}^+$ which is equal a.e. $/\mathfrak{M}$ to the Borel semicharacter k^∞, where $k^\infty(s) = 1$ if $h(s) = 1$ and $k^\infty(s) = 0$ if $h(s) < 1$.

The arc $\{h^r : 0 \leq r \leq \infty\}$ is degenerate exactly when h itself is an idempotent.

5.2.4. CRITICAL POINTS. An element of $\hat{S}^+$ is called a critical point if it cannot be approximated from below by strictly smaller elements of $\hat{S}^+$. That is, $h \in \hat{S}^+$ is critical if h is isolated (weakly) in $h\hat{S}^+ = \{f \in \hat{S}^+ : f \leq h\}$. Note that, by 5.1.5(a), we may replace weakly by strongly in this definition.

If $h \in \hat{S}^+$ is critical, then $h = h^2$ is an idempotent; otherwise, we would have $h^r < h$ for $r > 1$ and $\lim_{r \to 1+} h^r = h$.

PROPOSITION. *If $h \in \hat{S}^+$ then the following statements are equivalent*:

(a) *h is critical*;

(b) *the set $\Gamma_h = \{f \in S : |f| = h\}$ is strongly (weakly) open in $h\hat{S} = \{f \in \hat{S}$: $|f| \leq h\}$*;

(c) *h is a minimal element of some strongly open and closed set in $\hat{S}^+$*.

PROOF. The set Γ_h is the inverse image of $\{h\}$ under $f \to |f|$, while $h\hat{S}$ is the inverse image under $f \to |f|$ of $\{g \in \hat{S}^+ : g \leq f\}$. Since $f \to |f|$ is strongly continuous, (a) implies that Γ_h is strongly open in $h\hat{S}$. That this also forces Γ_h to be weakly open in $h\hat{S}$ follows from 5.1.5(a). Hence, (a) implies (b). Since $h = \hat{S}^+ \cap \Gamma_h$ and $hS = \hat{S}^+ \cap h\hat{S}$, we conclude that (b) implies (a) as well.

Now if h is critical, then h is a minimal element of the strongly open-closed subset of $\hat{S}^+$, $\{f \in \hat{S}^+ : h \leq f\} = \{f \in \hat{S}^+ : hf = h\}$; this is strongly open and closed because it is the inverse image of the isolated point $\{h\}$ under $f \to hf: \hat{S}^+ \to h\hat{S}^+$. Conversely, if h is a minimal element of some strongly open-closed set $\Omega \subset \hat{S}^+$, then $\{h\} = \Omega \cap \{f \in \hat{S}^+: f \leqslant h\}$ is isolated in $h\hat{S}^+ = \{f \in \hat{S}^+: f \leqslant h\}$ and h is critical.

The critical points of $\hat{S}^+$ play a major role in what follows. They determine the cohomology of $\hat{S}$ (Chapter 6) and arise from subalgebras of $\mathfrak{M}$ which are isomorphic to group algebras (Chapter 7).

5.2.5. COROLLARY. *If $h \in \hat{S}^+$ is a critical point, then its maximal group Γ_h is a locally compact topological group in the weak topology.*

PROOF. The group Γ_h is open in the closed set $h\hat{S}$ by 5.2.4.

§5.3. A covering lemma

5.3.1. Although $\hat{S}_\sigma$ may not be compact, there is a form of the covering property which holds in $\hat{S}_\sigma$ and substitutes for compactness in certain key arguments.

5.3.2. THE SETS $\Delta_h(\mu, \epsilon)$. If $h \in \hat{S}$ is an idempotent, we denote by Δ_h the closed ideal hS. If $0 \leqslant \mu \in \mathfrak{M}$ and $\epsilon > 0$ then $\Delta_h(\mu, \epsilon)$ will be the set $\{f \in \hat{S}: \int |f - hf|\, d\mu \leqslant \epsilon\} = \{f \in \hat{S}: \|f\mu - fh\mu\| \leqslant \epsilon\}$.

Note that $\Delta_h(\mu, \epsilon)$ is strongly closed and contains a strong neighborhood, $\{f \in \hat{S}: \|f\mu - fh\mu\| < \epsilon\}$, of Δ_h. In fact, since $h^2 = h$, the set Δ_h is exactly the set $\{f \in \hat{S}: f = fh\}$ and, thus, Δ_h is the intersection of the sets $\Delta_h(\mu, \epsilon)$ as μ runs over $\mathfrak{M}^+$ and ϵ runs over the positive reals.

5.3.3. PROPOSITION. *Each of the sets $\Delta_h(\mu, \epsilon)$ is a weakly closed ideal. Furthermore, every weak neighborhood of Δ_h contains $\Delta_h(\mu, \epsilon)$ for some μ and ϵ.*

PROOF. The statement $\int |f - fh|\, d\mu \leqslant \epsilon$ is equivalent to the statement that $|\hat{\nu}(f) - \hat{\nu}(fh)| = |\int (f - fh)\, d\nu| \leqslant \epsilon$ for every $\nu \in \mathfrak{M}^+$ with $\nu \leqslant \mu$. Hence, $\Delta_h(\mu, \epsilon)$ is indeed weakly closed.

If $f \in \Delta_h(\mu, \epsilon)$ and $g \in \hat{S}$, then $\int |fg - fgh|\, d\mu \leqslant \int |f - fh|\, d\mu \leqslant \epsilon$ and $fg \in \Delta_h(\mu, \epsilon)$. Thus, $\Delta_h(\mu, \epsilon)$ is an ideal.

The sets $\Delta_h(\mu, \epsilon)$, for fixed h, are directed downward under inclusion. In fact, $\Delta_h(\mu, \epsilon) \cap \Delta_h(\nu, \epsilon)$ contains $\Delta_h(\mu + \nu, \epsilon)$. Since $\bigcap_{\mu,\epsilon} \Delta_h(\mu, \epsilon) = \Delta_h$ and each $\Delta_h(\mu, \epsilon)$ is weakly compact, every weak niehgborhood of Δ_h must contain some $\Delta_h(\mu, \epsilon)$.

5.3.4. LEMMA. *Let A be a weakly closed ideal in $\hat{S}$, H a collection of idempotents in $\hat{S}$, $\{\mu_h : h \in H\}$ a collection of positive measures in $\mathfrak{M}$, and $\{\epsilon_h : h \in H\}$ a collection of positive numbers. If $A \subset \bigcup\{\Delta_h : h \in H\}$, then A is contained in the union of some finite subcollection of the sets $\Delta_h(\mu_h, \epsilon_h)$.*

PROOF. Since $f \in \Delta_h(\mu, \epsilon_h)$ if and only if $|f| \in \Delta_h(\mu_h, \epsilon_h)$, it suffices to prove that $\hat{S}^+ \cap A$ is covered by finitely many of these sets. If this were not true, we could choose a net $\{f_\alpha\} \subset \hat{S}^+ \cap A$ with the property that for each $h \in H$, $\{f_\alpha\}$ is eventually in the complement of $\Delta_h(\mu_h, \epsilon_h)$. Since $\hat{S}^+ \cap A$ is weakly compact, we could choose this net to be convergent in $\hat{S}^+ \cap A$. Thus, to prove the lemma, we need only show that a convergent net in $\hat{S}^+ \cap A$ is eventually in some $\Delta_h(\mu_h, \epsilon_h)$.

Thus, let $\{f_\alpha\} \subset \hat{S}^+ \cap A$ converge to $f \in \hat{S}^+ \cap A$ and choose $h \in H$ such that $f \in \Delta_h$. Then

$$\int |f_\alpha - f_\alpha h| \, d\mu_h = \int f_\alpha (1-h) \, d\mu_h = \int f_\alpha \, d\nu = \hat{\nu}(f_\alpha)$$

if $\nu = (1-h)\mu_h$. Also $\hat{\nu}(f_\alpha) \to \hat{\nu}(f) = \int (f - fh) \, d\mu_h = 0$. Hence, $f_\alpha \in \Delta_h(\mu_h, \epsilon_h)$ eventually, and the proof is complete.

CHAPTER 6. COHOMOLOGY OF $\hat{S}$

Although this chapter is devoted to characterizing the cohomology of $\hat{S}$, the reader needs no background in algebraic topology. Our methods apply to any tuple of functors from compact spaces to abelian groups which satisfies certain conditions. We shall verify directly that the pair of functors $\Delta \to H^0(\Delta)$, $\Delta \to H^1(\Delta)$ of Chapter 1 satisfies these conditions. It happens that Čech and sheaf cohomology do also (Bredon [1] and Spanier [1]).

§6.1. Cohomology functors

6.1.1. We shall be dealing with contravariant functors H from pairs $Y \subset X$ of compact Hausdorff spaces to abelian groups. That is, H assigns to each such pair $Y \subset X$ an abelian group $H(X, Y)$ and to each continuous map $\phi : (X, Y) \to (X', Y')$ ($\phi : X \to X'$ maps Y into Y') a group homomorphism $H(\phi) : H(X', Y') \to H(X, Y)$. Furthermore, H assigns the identity homomorphism to the identity map $\mathrm{id} : (X, Y) \to (X, Y)$ and H respects composition $(H(\phi \circ \psi) = H(\psi) \circ H(\phi))$.

We shall ordinarily denote $H(\phi)$ simply by ϕ^*.

For such a functor and a compact space X, we denote the group $H(X, \varnothing)$ by simply $H(X)$.

6.1.2. Excision. If $Y \subset X$ is a pair of compact Hausdorff spaces, we obtain a new pair $Y/Y \subset X/Y$ by identifying all points of Y. The space X/Y can be thought of as the one point compactification of $X \backslash Y$ with Y/Y as the point at infinity. The natural map $\pi : X \to X/Y$ is continuous, injective on $X \backslash Y$ and maps Y to the point Y/Y.

If H is a functor on pairs $Y \subset X$, as in 6.1.1, then $\pi : (X, Y) \to (X/Y, Y/Y)$ induces a map $\pi^* : H(X/Y, Y/Y) \to H(X, Y)$. We say that H is excisive if π^* is an isomorphism for any pair $Y \subset X$.

6.1.3. Continuity. If $K \subset X$ are compact spaces then the injection $i : K \to X$ induces a map $i^* : H(X) \to H(K)$. For $u \in H(X)$, we shall often refer to i^*u as the restriction of u to K. The map i^* will be called the restriction map and denoted by r_K.

We shall call H continuous if whenever K is the intersection of a family of compact sets K_α, with $\{K_\alpha\}$ directed downward, then (i) each element of $H(K)$ is the restriction of some element of $H(K_\alpha)$ for some α, and (ii) whenever $u \in H(X)$ satisfies $r_K u = 0$, it also satisfies $r_{K_\alpha} u = 0$ for some α.

The continuity condition simply says that $H(K)$ is the inductive limit of the directed system of groups $\{H(K_\alpha)\}$.

6.1.4. HOMOTOPY. Suppose we have a parameter space Ω, which is a Hausdorff topological space, and a family $\{\phi_\lambda\}_{\lambda \in \Omega}$ of maps from the compact space X to the compact space Y. The family is called continuous if the map $(x, \lambda) \to \phi_\lambda(x) : X \times \Omega \to Y$ is continuous.

Given such a family, there is a corresponding family of homomorphisms $\phi_\lambda^* : H(Y) \to H(X)$. We shall wish to deal with functors H which satisfy a strong version of the homotopy axiom: We shall require that any continuous family $\{\phi_\lambda\}$ have the property that $\phi_\lambda^* u$ is locally constant for each $u \in H(Y)$. Note that if Ω is connected, this implies that ϕ_λ^* is constant.

6.1.5. CONNECTED SEQUENCES, EXACTNESS. If H is a functor of the sort we have been discussing, and $Y \subset X$ are compact, then the injections $i : (Y, \varnothing) \to (X, \varnothing)$ and $j : (X, \varnothing) \to (X, Y)$ have the property that the composition $j \circ i : (Y, \varnothing) \to (X, Y)$ factors as $(Y, \varnothing) \to (Y, Y) \to (X, Y)$. If we assume that $H(Y, Y) = 0$ for each Y, then it follows that the composition $i^* \circ j^*$ in

$$H(X, Y) \xrightarrow{j^*} H(X) \xrightarrow{i^*} H(Y)$$

yields zero.

Now suppose we have a sequence $\{H^p\}_{p \in I}$ $(I = \{0, 1, \cdots, n\}$ or $I = \{0, 1, \cdots\})$ of such functors, and for each nonzero $p \in I$ and pair (X, Y) a map $\delta^{p-1} : H^{p-1}(Y) \to H^p(X, Y)$. We then call the system $\{H^p, \delta^p\}$ a connected sequence. We shall call it exact if for each pair (X, Y) the sequence

$$0 \longrightarrow H^0(X, Y) \xrightarrow{j^*} H^0(X) \xrightarrow{i^*} H^0(Y) \xrightarrow{\delta^0} H^1(X, Y)$$
$$\xrightarrow{j^*} \cdots \longrightarrow H^{p-1}(Y) \xrightarrow{\delta^{p-1}} H^p(X, Y) \xrightarrow{j^*} H^p(X) \xrightarrow{i^*} H^p(Y)$$

is exact for each $p \in I$. Note that we do not require exactness at the last stage of this sequence if $I = \{0, \cdots, n\}$ is finite and $p = n$.

6.1.6. COHOMOLOGY. By a cohomology sequence we shall mean an exact connected sequence $\{H^p\}_{p \in I}$, for which each H^p satisfies excision, continuity, and homotopy.

The sheaf and Čech cohomology theories for any coefficient group satisfy these conditions. In the next section, we shall prove that the pair of functors H^0 and H^1 of Chapter 1 can be interpreted as a (terminating) cohomology sequence.

6.1.7. LOCALLY COMPACT SPACES. Let X be locally compact and $X \cup \{\infty\}$ be its one point compactification. The injection $i : \{\infty\} \to X \cup \{\infty\}$ has a left inverse $k : X \cup \{\infty\} \to \{\infty\}$ which simply sends every point to infinity. Hence, for any cohomology sequence $\{H^p; \delta^p\}$, it follows that $i^* : H^p(X \cup \{\infty\}) \to H^p(\{\infty\})$ has a right inverse k^*. Thus, Ker (i^*) is a direct summand of $H^p(X \cup \{\infty\})$ and, by exactness, is isomorphic to $H^p(X \cup \{\infty\}, \{\infty\})$.

DEFINITION. We denote Ker $\{i^* : H^p(X \cup \{\infty\}) \to H^p(\{\infty\})\}$ by $H_c^p(X)$, and call it the pth cohomology group of X with compact supports.

§6.2. Idempotents and logarithms

6.2.1. THE FUNCTORS H^0 AND H^1. If (X, Y) is a pair of compact Hausdorff spaces with $Y \subset X$, we define $C(X, Y)$ to be the additive group of all continuous complex valued functions on X which vanish on Y. Similarly $C^{-1}(X, Y)$ is the multiplicative group of nonvanishing functions on X which are identically one on Y. If $\mathrm{ex}: C(X, Y) \to C^{-1}(X, Y)$ is the homomorphism defined by $\mathrm{ex}(f) = \exp(2\pi i f)$, then we define $H^0(X, Y)$ to be the kernel of ex and $H^1(X, Y)$ to be its cokernel.

Note that $H^0(X, Y)$ is the group of continuous integer valued functions on X which vanish on Y. In particular, $H^0(X) = H^0(X, \varnothing)$ is the group of all continuous integer valued functions on X and is the group generated by the idempotents of $C(X)$, as in Chapter 1.

Similarly, $H^1(X, Y)$ is the factor group of $C^{-1}(X, Y)$ modulo its subgroup consisting of elements which have continuous logarithms vanishing on Y. Thus, $H^1(X) = H^1(X, \varnothing)$ is the group $C(X)^{-1}/\exp(C(X))$ of Chapter 1.

6.2.2. THE CONNECTING HOMOMORPHISM. We define a map $\delta : H^0(Y) \to H^1(X, Y)$ as follows: If $f \in H^0(Y) \subset C(Y)$, we let $\tilde{f} \in C(X)$ be a continuous extension of f. Then we let δf be the equivalence class of $\mathrm{ex}(\tilde{f}) \in C^{-1}(X, Y)$ in the factor group $H^1(X, Y)$. Note that δf is independent of the choice of $\tilde{f}$ since the difference of two such extensions will be an element of $C(X, Y)$ and the image of $C(X, Y)$ under ex is the kernel of $C^{-1}(X, Y) \to H^1(X, Y)$.

6.2.3. PROPOSITION. *The connected sequence* $\{H^0, H^1, \delta\}$ *is exact. That is, for any pair* (X, Y) *the sequence*

(i) $$0 \longrightarrow H^0(X, Y) \xrightarrow{j^*} H^0(X) \xrightarrow{i^*} H^0(Y) \xrightarrow{\delta} H^1(X, Y) \xrightarrow{j^*} H^1(X) \xrightarrow{i^*} H^1(Y)$$

is exact.

PROOF. This is a standard chase through the commutative diagram

$$\begin{array}{ccccccc} 0 \longrightarrow & C(X, Y) & \xrightarrow{j^*} & C(X) & \xrightarrow{i^*} & C(Y) & \longrightarrow 0 \\ & \downarrow \mathrm{ex} & & \downarrow \mathrm{ex} & & \downarrow \mathrm{ex} & \\ 0 \longrightarrow & C^{-1}(X, Y) & \xrightarrow{j^*} & C^{-1}(X) & \xrightarrow{i^*} & C^{-1}(Y). & \end{array}$$

Here the map j^* induced by $j : (X, \varnothing) \to (X, Y)$ is just the inclusion map of $C(X, Y)$ $(C^{-1}(X, Y))$ into $C(X)$ $(C^{-1}(X))$, while the map i^* induced by $i : (Y, \varnothing) \to (X, \varnothing)$ is just restriction to Y of functions on X. Note that the kernel of i^* is exactly j^* and that j^* is injective on each row. That i^* is surjective on the top row is just the Tietze extension theorem.

The map δ is defined by the condition that δf is the equivalence class containing the set $j^{*-1} \circ \mathrm{ex} \circ i^{*-1}(f) \subset C^{-1}(X, Y)$.

Verifying that (i) is exact is now a tedious but quite standard and routine diagram chase.

6.2.4. EXCISION. We claim that H^0 and H^1 satisfy excision. Thus, let (X, Y) be a pair and $\pi:(X, Y) \to (X/Y, Y/Y)$ the map which identifies Y to a point. The map $f \to f \circ \pi$ is an isomorphism from $C(X/Y, Y/Y)$ to $C(X, Y)$ and from $C^{-1}(X/Y, Y/Y)$ to $C^{-1}(X, Y)$, and it commutes with ex. It follows that this map induces isomorphisms $\pi^*: H^0(X/Y, Y/Y) \to H^0(X, Y)$ and $\pi^*: H^1(X/Y, Y/Y) \to H^1(X, Y)$ on the kernel and cokernel, respectively, of ex.

6.2.5. CONTINUITY. Suppose $\{Y_\alpha\}$ is a family of compact subsets of X, which is directed downward and has intersection Y.

If $f \in C^{-1}(Y)$, then it has a continuous extension to an element $\tilde{f} \in C(X)$, which necessarily is nonvanishing in some neighborhood of Y. Since some Y_α will be contained in this neighborhood, f is the restriction of an element of $C^{-1}(Y_\alpha)$ for some α. Trivially, every $g \in C(Y)$ has an extension in $C(X)$ and, hence, in each $C(Y_\alpha)$. Now if $f \in C^{-1}(Y)$ has a logarithm $g \in C(Y)$ and $\tilde{f}, \tilde{g}$ are extensions of f, g in $C^{-1}(Y_\alpha)$, $C(Y_\alpha)$ for some α, then $\tilde{f}\,\mathrm{ex}(-\tilde{g})$ is one on Y. Thus, in any neighborhood of Y on which $|1 - \tilde{f}\,\mathrm{ex}(-\tilde{g})| < 1$ there is a continuous $\tilde{h}$ for which $\mathrm{ex}(\tilde{h}) = \tilde{f}\,\mathrm{ex}(-\tilde{g})$; i.e., $\tilde{f} = \mathrm{ex}(\tilde{h} + \tilde{g})$. It follows that f is the restriction of some element of $\mathrm{ex}(C(Y_\alpha))$ for sufficiently large α.

The above considerations obviously imply that H^1 satisfies continuity. That H^0 also satisfies continuity follows from similar considerations. In fact, if $h \in H_0(Y)$ and $\tilde{h} \in C(X)$ is any extension of h, the function $\mathrm{ex}(\tilde{h})$ is one on Y and, therefore, has a logarithm $\tilde{g}$ in some neighborhood V of Y which is zero on Y. Then $\mathrm{ex}(\tilde{h} - \tilde{g}) = 1$ on V and $\tilde{h} = \tilde{h} - \tilde{g}$ on Y. It follows that $\tilde{h} - \tilde{g} \in H^0(Y_\alpha)$ for any $Y_\alpha \subset V$ and is an extension of h. Clearly, if h vanishes on Y and is integer valued, it vanishes on a neighborhood of Y, hence on some Y_α.

6.2.6. HOMOTOPY. Let X and Y be compact spaces and $\{\phi_\lambda\}_{\lambda\in\Omega}$ $(\phi_\lambda : X \to Y)$ a continuous family of maps over the topological space Ω.

Suppose $f \in C(Y)$ and let $f' \in C(X \times \Omega)$ be defined by $f'(x, \lambda) = f(\phi_\lambda(x))$. Note that if f is integer valued, so is f'; if f is nonvanishing, so is f'; and if $f = \mathrm{ex}(g)$ then $f' = \mathrm{ex}(g')$. Note also, that for each $\lambda \in \Omega$, $\phi_\lambda^* f(x) = f'(x, \lambda)$.

Now if $h \in H^0(X)$ and $h = \phi_{\lambda_0}^* f$ for some $\lambda_0 \in \Omega$, $f \in H^0(Y)$, we have that the function $(x, \lambda) \to f(x, \lambda) - h(x)$ vanishes on $X \times \{\lambda_0\}$. Since X is compact and h and f' are integer valued and continuous on X and $X \times \Omega$, this identity must hold on $X \times U$ for some neighborhood U of λ_0. It follows that $h = \phi_\lambda^* f$ for all $\lambda \in U$.

Similarly, if $h \in C^{-1}(X), f \in C^{-1}(Y)$ and $\phi_{\lambda_0}^* [f] = [h]$ holds for the classes $[f] \in H^1(Y)$ and $[h] \in H^1(X)$, then $f'(x, \lambda_0)\,h(x)^{-1} = \mathrm{ex}(g(x))$ for some $g \in C(X)$. That is, $(x, \lambda) \to f'(x, \lambda)h(x)^{-1}\,\mathrm{ex}(-g(x))$ is one on $X \times \{\lambda_0\}$. It follows that this function has a logarithm in a neighborhood $X \times U$ of $X \times \{\lambda_0\}$. We conclude that $\phi_\lambda^*[f] = [h]$ for all $\lambda \in U$.

Thus, we have verified that H^0 and H^1 satisfy the homotopy axiom of 6.1.4.

Combining the results of this section, we have:

6.2.7. THEOREM. *The connected sequence* $\{H^0, H^1; \delta\}$ *is a cohomology sequence in the sense of* §6.1.

§6.3. The cohomology of $\hat{S}$

6.3.1. We now assume that S is the structure semigroup and $\hat{S}$ the spectrum of a commutative, semisimple measure algebra $\mathfrak{M}$ with a norm one identity. We use the results of Chapter 5 and the axioms of cohomology to compute the groups $H^p(\hat{S})$ for any cohomology sequence $\{H^p; \delta^p\}$.

6.3.2. AN APPLICATION OF HOMOTOPY. Recall that $(h, f) \to hf: S_\sigma \times S_\omega \to S_\omega$ is continuous (5.1.4). Thus, we may define a continuous family of maps $\{\phi_h\}_{h \in \hat{S}_\sigma}$ from $\hat{S}_\omega$ to $\hat{S}_\omega$ by $\phi_h(f) = hf$. We shall actually only be interested in these maps for $h \in \hat{S}^+$.

Now if $A \subset \hat{S}$ is a weakly closed (hence, compact) ideal, then the maps ϕ_h, for $h \in \hat{S}^+$, leave A invariant. Hence, we may also consider $\{\phi_h\}_{h \in \hat{S}}$ to be a continuous family of maps of A into A.

For each closed ideal A, the map $\phi_h^*: H^p(A) \to H^p(A)$ induced on cohomology by ϕ_h will be denoted π_h. For an ideal $B \subset A$, the restriction map $H^p(A) \to H^p(B)$ induced by $B \to A$ will be denoted by r_B.

PROPOSITION. *For closed ideals* $B \subset A \subset \hat{S}$, *the family* $\{\pi_h\}_{h \in \hat{S}^+}$ *is a family of endomorphisms of* $H^p(A)$ $(H^p(B))$ *with the following properties*:

(a) $\pi_h \circ \pi_g = \pi_{hg}$ *for* $h, g \in \hat{S}^+$;

(b) *for each* $u \in H^p(A)$, $h \to \pi_h u$ *is locally constant on* $\hat{S}_\sigma^+$;

(c) *each* π_h *commutes with the restriction map* r_B;

(d) *each* π_h *is a projection* $(\pi_h^2 = \pi_h)$ *onto a subgroup of* $H^p(A)$; *furthermore, the restriction map* $r_{hA}: H^p(A) \to H^p(hA)$ *maps this subgroup isomorphically onto* $H^p(hA)$.

PROOF. Part (a) follows from the fact that $\phi_{hg} = \phi_h \circ \phi_g$. Part (b) is a consequence of the homotopy axiom. Part (c) is due to the commutativity of the diagram

$$\begin{array}{ccc} B & \xrightarrow{\;i\;} & A \\ \downarrow{\scriptstyle \phi_h} & & \downarrow{\scriptstyle \phi_h} \\ B & \xrightarrow{\;i\;} & A \end{array}$$

where i is the injection.

That $\pi_h^2 = \pi_h$ is clear if h is an idempotent. However, each $h \in \hat{S}^+$ is on an arc $(h^r, 0 \leqslant r \leqslant \infty)$ in $\hat{S}^+$ with idempotents as endpoints. Hence, by (b), $\pi_h^2 = \pi_h$ holds for every $h \in \hat{S}^+$.

To complete the proof of part (d), we must show that $r_{hA}: H^p(A) \to H^p(hA)$ is an isomorphism of $\operatorname{Im}(\pi_h) \subset H^p(A)$ onto $H^p(hA)$. Let $\gamma_h: H^p(hA) \to H^p(A)$ denote the map induced on cohomology by $f \to hf: A \to hA$. Then $\pi_h = \gamma_h \circ r_{hA}$, since ϕ_h is $f \to hf: A \to hA$, followed by $i: hA \to A$. On the other hand, if h is idempotent, then $r_{hA} \circ \gamma_h = \mathrm{id}: H^p(hA) \to H^p(hA)$, since $f \to hf$ is then the identity on hA. It follows that γ_h is a right inverse for r_{hA} with image equal to $\mathrm{Im}(\pi_h)$—at least when h is idempotent. The fact that this holds for all $h \in \hat{S}^+$ now follows from the homotopy axiom and the fact that each such h is connected by an arc to an idempotent. The maps γ_h, π_h, and the group $H^p(hA)$ remain fixed along such an arc.

6.3.3. A THEOREM ON IDEALS. The main theorem of this section has three interrelated parts which will be proved by induction on the dimension.

THEOREM. *Let* $\{H^p, \delta^p\}$ *be a cohomology sequence and* $A \subset \hat{S}$ *a closed ideal. Then*:

(a) *if* $A = A_1 \cup A_2 \cup \cdots \cup A_n$ *with each* A_i *a closed ideal and if* $u \in H^p(A)$ *satisfies* $r_{A_i} u = 0$ *for each* i, *then* $u = 0$;

(b) *each* $u \in H^p(A)$ *is generated by the elements* $\pi_h u$ *for* $h \in A \cap \hat{S}^+$;

(c) *the restriction* $r_A : H^p(\hat{S}) \to H^p(A)$ *is surjective.*

The proof will proceed as follows: we first show that for a given p, (a) implies (b) and (b) implies (c). We then show that if (c) holds for $p-1$ (or if $p=0$) then (a) holds for p. If we do this, then the theorem follows by induction.

6.3.4. LEMMA. *If* (a) *of Theorem* 6.3.3 *holds for a given* p, *then so does* (b).

If $h \in A \cap \hat{S}^+$ and $k = h^0$, then $\pi_h u = \pi_k u$ by 6.3.2(b). Although k may no longer be an element of $A \cap \hat{S}^+$, we can conclude that to prove 6.3.3(b) it suffices to prove that u is generated by the set of elements $\pi_k u$ for which k is an idempotent of the form $k = h^0$ with $h \in A \cap \hat{S}^+$. We denote the collection of such idempotents by K.

Note that, for $k \in K$, $f \to kf$ is the identity on kA. It follows that $r_{kA}\pi_k u = \pi_k r_{kA} u = r_{kA} u$ and, hence, that $u - \pi_k u$ restricts to zero on kA.

By the continuity axiom (6.1.3) and Proposition 5.3.3, there is, for each k, an ideal of the form $\Delta_k(\mu_k, \epsilon_k)$ such that $u - \pi_k u$ restricts to zero on $A \cap \Delta_k(\mu_k, \epsilon_k)$. By Lemma 5.3.4, there are finitely many of these ideals, say $\Delta_{k_1}(\mu_{k_1}, \epsilon_{k_1}), \cdots, \Delta_{k_n}(\mu_{k_n}, \epsilon_{k_n})$, which cover A. We set $A_i = \Delta_{k_i}(\mu_{k_i}, \epsilon_{k_i})$. Then, for each i, $r_{A_i}(1 - \pi_{k_i})u = 0$.

We now let v be the element

$$v = \left[1 - \prod_{i=1}^{n} (1 - \pi_{k_i})\right] u$$

and observe, by expanding the product, that v is an alternating sum of elements of the form $\pi_k u$ with k a product of elements of the set K. Since K is closed under products, v is in the subgroup of $H^p(A)$ generated by elements $\pi_k u$ with $k \in K$. Furthermore,

$$r_{A_j}(u - v) = r_{A_j} \prod_{i=1}^{n} (1 - \pi_{k_i})u = 0,$$

since $r_{A_j}(1 - \pi_{k_j})u = 0$. (This uses the commutativity observed in 6.3.2(c).) Thus, if (a) of Theorem 6.3.3 holds, then $u = v$.

6.3.5. LEMMA. *If* (b) *of Theorem* 6.3.3 *holds for a given* p, *then so does* (c) *of Theorem* 6.3.3.

PROOF. If $u \in H^p(A)$ is in the image of $\pi_h : H^p(A) \to H^p(A)$ for some $h \in A \cap \hat{S}^+$, then u is also in the image of $r_A : H^p(\hat{S}) \to H^p(A)$. In fact, by 6.3.2(d), r_{hA} is an isomorphism onto $H^p(hA)$ when restricted to the image of π_h in either $H^p(\hat{S})$ or $H^p(A)$. Thus, r_A must map the image of π_h in $H^p(\hat{S})$ onto the image of π_h in $H^p(A)$.

Since $H^p(A)$ is generated by elements u of the above form if we assume (b) of 6.3.3, we conclude that $r_A : H^p(\hat{S}) \to H^p(A)$ is surjective in this case.

6.3.6. LEMMA. *If* $p = 0$ *or if* $p > 0$ *and* (c) *of* 6.3.3 *holds for* $p - 1$, *then* (a) *of* 6.3.3 *holds for* p.

PROOF. This is where we use the exactness axiom (6.1.4). Note that, under the hypothesis of the lemma, it implies that for a pair $B \subset A$ of closed ideals, the sequence

(i) $$0 \longrightarrow H^p(A, B) \xrightarrow{j^*} H^p(A) \xrightarrow{r_B} H^p(B)$$

is exact. For $p = 0$ this is clear; for $p > 0$ it follows from the fact that $H^{p-1}(A) \to H^{p-1}(B)$ is surjective (which follows from 6.3.3(c) for $p - 1$).

It is clearly sufficient to prove 6.3.3(a) in the case $n = 2$, since the general case then follows by induction (the union of finitely many closed ideals is also a closed ideal). Hence, let $A = B \cup C$ with B, C closed subideals of A, and consider the commutative diagram:

$$\begin{array}{ccccccc} 0 & \longrightarrow & H^p(A, B) & \xrightarrow{j^*} & H^p(A) & \xrightarrow{r_B} & H^p(B) \\ & & \downarrow r_C & & \downarrow r_C & & \downarrow r_{B\cap C} \\ 0 & \longrightarrow & H^p(C, B \cap C) & \xrightarrow{j^*} & H^p(C) & \xrightarrow{r_{B\cap C}} & H^p(B \cap C). \end{array}$$

If $u \in H^p(A)$ and $r_B u = r_C u = 0$, then $u = j^* v$ for a unique $v \in H^p(A, B)$. Furthermore, $j^* r_C v = r_C j^* v = r_C u = 0$. It follows from the exactness of the bottom row of (ii) (which is (i) with A and B replaced by C and $B \cap C$) that $r_C v = 0$. However, it follows from excision (6.1.2) that r_C is an isomorphism from $H^p(A, B)$ to $H^p(C, B \cap C)$. Hence, $v = 0$ and $u = 0$.

This completes the proof of Theorem 6.3.3.

§6.4. Cohomology and critical points

6.4.1. THE SUBGROUPS J_h^p. For each critical point $h \in \hat{S}^+$, we set $A_h = hS = \{f \in \hat{S} : |f| \leqslant h\}$ and $B_h = \{f \in \hat{S} : |f| < h\}$. Note that A_h and B_h are ideals, A_h is closed, and (by 5.2.4) B_h is closed. The maximal group at h, $\Gamma_h = \{f \in \hat{S} : |f| = h\}$, is $A_h \backslash B_h$.

We denote the subgroup $\mathrm{Im}(\pi_h) \cap \mathrm{Ker}(r_{B_h})$ by J_h^p.

PROPOSITION. *The group* J_h^p *consists of those* $u \in H^p(\hat{S})$ *for which* $\pi_g u = u$ *if* $g \geqslant h$ *and* $\pi_g u = 0$ *otherwise.*

PROOF. If $g \geqslant h$ then $gh = h$ and $\pi_g \circ \pi_h = \pi_h$; hence, $\pi_g u = u$ for all $u \in \mathrm{Im}(\pi_h)$. If $g \ngeqslant h$, then $gh < h$ and, thus, $gh \in B_h$. Since r_{B_h} is then an isomorphism on $\mathrm{Im}(\pi_{gh})$ and $r_{B_h} \pi_{gh} u = r_{B_h} \pi_g u = \pi_g r_{B_h} u = 0$ for $u \in J_h^p$, we have that π_g vanishes on J_h^p. Conversely, if $\pi_g u = 0$ for all $g < h$, then $\pi_g r_{B_h} u = 0$ for all $g \in B_h \cap \hat{S}^+$. We then use 6.3.3(b) to conclude that $u \in \operatorname{Ker}(r_{B_h})$.

Now if $\theta_h : \hat{S} \to \Gamma_h \cap \{\infty\}$ is the map $f \to hf : \hat{S} \to A_h$ followed by the identification map $A_h \to A_h/B_h = \Gamma_h \cup \{\infty\}$, then

6.4.2. PROPOSITION. *The map* $\theta_h^* : H^p(\Gamma_h \cup \{\infty\}) \to H^p(\hat{S})$, *when restricted to* $H_c^p(\Gamma_h)$, *is an isomorphism of* $H_c^p(\Gamma_h)$ *onto* J_h^p.

PROOF. The restriction map $r_{A_h} : H^p(\hat{S}) \to H^p(A_h)$ maps J_h^p isomorphically onto the kernel of $r_{B_h} : H^p(A_h) \to H^p(B_h)$ (6.3.2(a)). By exactness, this is exactly the image in $H^p(A_h)$ of $j^* : H^p(A_h, B_h) \to H^p(A_h)$; furthermore, j^* is injective since $r_{B_h} : H^{p-1}(A_h) \to H^{p-1}(B_h)$ is surjective. The proposition now follows from the excision axiom (since A_h/B_h is $\Gamma_h \cup \{\infty\}$) and the fact that $H_c^p(\Gamma_h)$ is isomorphic to $H^p(\Gamma_h \cup \{\infty\}, \{\infty\})$ (6.1.7).

Now it is not difficult to compute the cohomology of the one point compactification of an l. c. a. group such as Γ_h. In the next chapter we shall show how to identify the critical points h and the groups Γ_h for any measure algebra $\mathfrak{M}$. In view of the next theorem, this will allow us to compute $H^p(\hat{S})$ for any measure algebra.

6.4.3. THEOREM. *For each* p, $H^p(\hat{S})$ *is the direct sum of its subgroups* J_h^p *as* h *ranges over the cirtical points of* $\hat{S}^+$

We shall prove this in two steps. We first prove that the J_h^p's generate $H^p(\hat{S})$ and then that they are independent.

6.4.4. LEMMA. *The groups* J_h^p *(*h *a critical point) generate* $H^p(\hat{S})$.

PROOF. For a given $u \in H^p(\hat{S})$, consider the set Ω of all $g \in \hat{S}^+$ for which $\pi_g u$ is not an element of the subgroup K^p of $H^p(\hat{S})$ generated by the groups J_h^p. Since $g \to \pi_g u$ is locally constant on $\hat{S}_\sigma^+$, the set Ω is both open and closed in $\hat{S}_\sigma^+$. We wish to show that it is empty. In fact if $1 \notin \Omega$ then $u = \pi_1 u$ is in the subgroup K^p.

Suppose Ω is not empty. Then by 5.1.6 it has a minimal element h, which must be a critical point by 5.2.4. Thus, $\pi_h u \notin K^p$, but $\pi_g u \in K^p$ for all $g \in \hat{S}^+$ with $g < h$.

By Theorem 6.3.3(b), $r_{B_h} u = w$ is generated by elements of the form $\pi_g w$ for $g \in B_h \cap \hat{S}^+$, i.e., for $g \in \hat{S}^+$ with $g < h$. Hence, we conclude that $r_{B_h} u = r_{B_h} v$ for some $v \in K^p$. But then $r_{B_h}(u - v) = 0$ and hence $\pi_h(u - v) \in J_h^p$. By 6.4.1, $\pi_h u \in K^p$ if v is and, thus, $\pi_h u \in K^p$. This contradicts the choice of h and proves that $\Omega = \emptyset$. Hence, $K^p = H^p(\hat{S})$.

6.4.5. LEMMA. *Let* $h_1, \cdots, h_k$ *be a set of distinct critical points, and* u_i *a non-zero element of* $J_{h_i}^p$ *for each* i. *If* $\Sigma_{i=1}^k u_i = 0$, *then* $u_1 = u_2 = \cdots = u_k = 0$.

PROOF. Suppose the contrary and let k be the smallest integer for which a sum $\Sigma_{i=1}^k u_i$ with $0 \neq u_i \in J_{h_i}^p$ for distinct $h_1, \cdots, h_n$ can be zero. For such a sum, let Ω be the set of all $g \in S^+$ for which $\pi_h u_1 = u_1$. This set is open and closed in $\hat{S}_\sigma^+$ and, hence, has a minimal element h which is a critical point. Thus $\pi_h u_1 = u_1$ and $\pi_g u_1 = 0$

for $g < h$. However, this implies that for each i, $\pi_h u_i = u_i$ and $\pi_g u_i = 0$ for g/h by the choice of k. (Recall either $\pi_g u_i = 0$ or $\pi_g u_i = u_i$.) It follows that each u_i is an element of J_h^p. However, we deduce from 6.4.1 that the groups J_h^p and $J_{h_i}^p$ meet only in zero if $h \neq h_i$. Thus we have contradicted the hypothesis that the h_i's were distinct.

CHAPTER 7. CRITICAL POINTS AND GROUP ALGEBRAS

In this chapter, $\mathfrak{M}$ will again be a semisimple, commutative measure algebra with structure semigroup S and spectrum $\hat{S}$. As before, we consider $\mathfrak{M}$ to be an L-subalgebra of $M(S)$. We establish a one-to-one correspondence between (maximal) group algebras contained in $\mathfrak{M}$ and critical points of $\hat{S}^+$.

§7.1. The main theorem

7.1.1. If $h \in \hat{S}^+$ is any idempotent, then $N_h = \{s \in S : h(s) = 0\}$ is an open-closed ideal of S and $S_h = \{s \in S : h(s) = 1\} = S \backslash N_h$ is an open-closed subsemigroup of S. In the terminology of 3.5.4, N_h is a prime ideal of S.

Now the kernel of the semigroup S_h is a maximal group in S. We denote it by K_h. We are interested in when a maximal group $G \subset S$ has this form, and particularly, when it has this form for h a critical point.

Proposition. *If $G \subset S$ is a maximal group which is not a set of measure zero for all $\mu \in \mathfrak{M}$, then $G = K_h$ for a critical point $h \in \hat{S}^+$.*

Proof. If G is not a set of measure zero for each $\mu \in \mathfrak{M}$, then we can choose $\mu \in \mathfrak{M}$ such that μ is positive, of norm one. and supported on G.

Now the restriction to G of an element of $\hat{S}$ is either a character of G or is identically zero. Hence, each $f \in \hat{S}^+$ is either identically one or identically zero on G. Thus, for μ above, $\hat{\mu}(f) = 0$ or $\hat{\mu}(f) = 1$ for each $f \in \hat{S}^+$. It follows that $\Omega = \{f \in \hat{S}^+ : \hat{\mu}(f) = 1\} = \{f \in \hat{S}^+ : f|_G = 1\}$ is both open and closed in $\hat{S}^+$. By 5.1.6 and 5.2.4 it has a minimal element h which is a critical point.

To complete the proof, we must show that $G = K_h$. Clearly, $G \subset S_h = \{s \in S : h(s) = 1\}$. Thus, if p and q are the idempotents in $K_h = \text{Kernel}(S_h)$ and G, respectively, then $pq = p$. If p and q are different, there is some $f \in \hat{S}$ such that $f(p) \neq f(q)$ (by 3.2.3). Now the equations $p^2 = p$, $q^2 = q$, and $pq = p$ imply that $f(p) = 0$ and $f(q) = 1$, from which it follows that $|f| < h$ and $|f|h \equiv 1$ on G. This contradicts the minimality of h and establishes that $p = q$. Hence, $G \subset K_h$; the maximality of G then implies that $G = K_h$.

The next proposition describes one way (it turns out to be the only way) that maximal groups supporting nonzero measures in $\mathfrak{M}$ arise.

7.1.2. Proposition. *If $\mathfrak{N}$ is an L-subalgebra of $\mathfrak{M}$ which is isomorphic (as a measure algebra) to the group algebra of some l. c. a. group, then the support of $\mathfrak{N}$ is a* *subgroup of some maximal group in* S. *Necessarily* (*by* 7.1.1) *this maximal group is* K_h *for some critical point* h.

PROOF. If $\mathfrak{N} \simeq L(G)$ for an l. c. a. group G, then by 3.3.4 the map $L(G) \to \mathfrak{N} \subset \mathfrak{M}$ is induced by a continuous isomorphism of $\overline{G}$, the structure semigroup of $L(G)$, into S. The image of α is a group in S and is the support of $\mathfrak{N}$. This group is contained in a maximal group which must be K_h for some critical point h by 7.1.1.

The main theorem of this chapter amounts to a converse for 7.1.2.

7.1.3. THEOREM. *If* h *is a critical point of* $\hat{S}$, K_h *is the kernel of* $S_h = \{s \in S : h(s) = 1\}$, *and* Γ_h *the locally compact group* $\{f \in \hat{S} : |f| = h\}$, *then*

(a) Γ_h *is the dual group and* K_h *the Bohr compactification of an l. c. a. group* G_h;

(b) $L(G_h)$ *is isomorphic to an* L-*subalgebra of* $\mathfrak{M}$ *with support* K_h, *where the isomorphism is the restriction to* $L(G_h)$ *of the map* $\mu \to \mu \circ \alpha^{-1} : M(G_h) \to M(K_h)$ *induced by the natural map* $\alpha : G_h \to K_h$ *of* G_h *into its Bohr compactification*;

(c) *there is an* L-*subalgebra* $\mathfrak{N}$ *of* $M(G_h)$ *such that* $L(G_h) \subset \mathfrak{N} \subset \mathrm{Rad}(L(G_h))$ *and* $\mu \to \mu \circ \alpha^{-1}$ *maps* $\mathfrak{N}$ *onto* $\mathfrak{M} \cap M(K_h)$.

The proof of this theorem will occupy most of the remainder of this chapter. The construction of the group G_h is no mystery: it is simply the dual group of the l. c. a. group Γ_h. By the Pontryagin duality theorem (Rudin [5]), we may identify Γ_h with the dual group of G_h. Note that K_h is a compact group having Γ_h (with the discrete topology) as its dual group. Hence, K_h is the Bohr compactification of G_h (4.1.2). However, it is quite difficult to show that K_h supports nonzero measures in $\mathfrak{M}$ and even more difficult to show that some of these are images of absolutely continuous measures on G_h.

7.1.4. REMARK. It is sufficient to prove the above theorem in the case where the critical point in question is the identity. In fact, any critical point h agrees with the identity on the set $S_h = \{s \in S : h(s) = 1\}$. Furthermore, S_h is the structure semigroup of the L-subalgebra $\mathfrak{M}_h$ of $\mathfrak{M}$ consisting of measures concentrated on S_h (3.5.3). The spectrum $\hat{S}_h$ of $\mathfrak{M}_h$ may be identified with $h\hat{S}$ since the map $f \to fh$ is a retract of $\hat{S}$ onto $\hat{S}_h$ and the Gelfand topology induced by $\mathfrak{M}$ on $\hat{S}$ agrees on $h\hat{S}$ with the topology induced by $\mathfrak{M}_h$. Hence, the identity h of $h\hat{S} = \hat{S}_h$ is a critical point for the algebra $\mathfrak{M}_h$.

Thus, Theorem 7.1.3 will follow if we can prove

THEOREM. *If* $\mathfrak{M}$ *is a commutative, semisimple measure algebra such that* 1 *is a critical point of* $\hat{S}$, *then there is an l. c. a. group* G *and an* L-*subalgebra* $\mathfrak{N} \subset M(G)$ *such that*

(a) $L(G) \subset \mathfrak{N} \subset \mathrm{Rad}\, L(G)$;

(b) *the kernel* K *of* S *is the Bohr compactification of* G *and* $\Gamma = \{f \in \hat{S} : |f| = 1\}$ *is the dual group of* G;

(c) *the natural embedding* $G \to K$ *induces an isomorphism of* $\mathfrak{N}$ *onto* $\mathfrak{M} \cap M(K)$.

In proving this theorem in the ensuing sections, we shall assume that $\mathfrak{M}$ has a normalized identity. We can do this without loss of generality since adjoining such an identity to $\mathfrak{M}$ does not affect either $\mathfrak{M} \cap M(K)$ or the fact that 1 is a critical point.

7.1.5. MAXIMAL GROUP ALGEBRAS. If L is any L-subalgebra of $\mathfrak{M}$ which is isomorphic to a group algebra, then, by 7.1.2, L is supported on some maximal group K_h for h a critical point. Let L_h be the image of $L(G_h)$ in $\mathfrak{N}_h = \mathfrak{M} \cap M(K_h)$, as in 7.1.3. Now $\mathfrak{N}_h / L_h$ is isomorphic to a subalgebra of Rad $L(G_h)/L(G_h)$ and, hence, is a radical algebra. It follows that $L_h \cap L \neq (0)$ since, otherwise, L would be a radical algebra. However, L_h is an L-ideal of $\mathfrak{N}_h$ and, hence, $L_h \cap L$ is an L-ideal of L. Since a group algebra contains no nonzero proper L-ideals, we conclude that $L \subset L_h$. Hence,

PROPOSITION. *For each critical point h the image L_h of $L(G_h)$ in $\mathfrak{M} \cap M(K_h)$ is maximal among L-subalgebras of $\mathfrak{M}$ which are isomorphic to group algebras. Furthermore, each L-subalgebra of $\mathfrak{M}$ which is isomorphic to a group algebra is contained in L_h for some h.*

7.1.6. COHOMOLOGY. Given a Banach algebra A with identity and a closed subalgebra B (which may not contain the identity), the restriction map $f \to f|_B : \Delta(A) \to \Delta(B) \cup \{0\}$ is a continuous map from the spectrum of A to the one-point compactification of the spectrum of B. Hence, it induces a homomorphism $H_c^p(\Delta(B)) \to H^p(\Delta(A))$ of cohomology for every cohomology sequence. We shall call this the canonical map.

If we assume 7.1.3, then Theorem 6.4.3 and Proposition 6.4.2 can be restated as follows.

THEOREM. *Let $\mathfrak{M}$ be a commutative, semisimple measure algebra with normalized identity and with spectrum $\hat{S}$, and let $\{H^p\}$ be any cohomology sequence. Then for each maximal group algebra $L_h \subset \mathfrak{M}$ with spectrum Γ_h, the canonical map of $H_c^p(\Gamma_h)$ into $H^p(\hat{S})$ is injective. Furthermore, $H^p(\hat{S})$ is the direct sum of the images of these maps as L_h ranges over all maximal group algebras (h ranges over all critical points).*

§7.2. Linear equations and absolutely continuous measures in $M(G_0 \times R^n)$

7.2.1. It is apparent that to prove Theorem 7.1.4 we must have techniques for constructing measures in $\mathfrak{M}$ which are images of absolutely continuous measures on G. In this section we shall present a technique for constructing absolutely continuous measures in $M(G)$ for a group of the form $G = G_0 \times R^n$, where G_0 is compact. This technique is the core of the proof of 7.1.4.

Since G has R^n as a subgroup, we shall use additive notation for the group operation in G.

7.2.2. THE SEMIGROUPS A AND A_j. We let A and A_j, for $j = 1, \cdots, n$, denote the subsemigroups of G defined by

$$A = G_0 \times \{x = (x_1, \cdots, x_n) \in R^n : x_i \geqslant 0 \quad \text{for} \quad i = 1, \cdots, n\},$$

$$A_j = G_0 \times \{x \in R^n : x_i \geqslant 0 \quad \text{for} \quad i \neq j\}.$$

Note that $G = A_1 + A_2 + \cdots + A_n$, but the sum $\Sigma_{j \neq i} A_j$ is the half space

$$H_i = G_0 \times \{x \in R^n : x_i \geqslant 0\}.$$

The half spaces H_i are all disjoint from the set

$$B = -A^\circ = G_0 \times \{x \in R^n : x_i < 0 \text{ for } i = 1, \cdots, n\}.$$

We consider each of the algebras $M(A)$ and $M(A_i)$ to be a subalgebra of $M(G)$.

7.2.3. THE RESIDUE MEASURE. Let $\mu_1, \cdots, \mu_n \in M(A)$ be measures for which the equation

(i) $$\mu_1 * \nu_{j1} + \cdots + \mu_n * \nu_{jn} = \delta$$

has a solution $\nu_{j1}, \cdots, \nu_{jn} \in M(A_j)$ for each j. By a residue measure for $(\mu_1, \cdots, \mu_n)$ we mean a measure of the form $\rho = \det(\nu_{ij}) \in M(G)$ for any set of solutions to (i).

A residue measure for $(\mu_1, \cdots, \mu_n)$ is not unique; however, a certain portion of it is.

PROPOSITION. *If ρ and ρ' are residue measures for $(\mu_1, \cdots, \mu_n)$, then $(\rho - \rho')|_B = 0$.*

PROOF. It suffices to prove this in the case where ρ' is obtained from $\rho = \det(\nu_{ij})$ by modifying one row of the matrix (ν_{ij}). Without loss of generality, we may assume this is the first row.

Thus, let ρ' be obtained by replacing $\nu_{11}, \cdots, \nu_{1n}$ by a solution $\nu'_{11}, \cdots, \nu'_{1n} \in M(A_1)$ for (i). Then the following equation is satisfied:

(ii) $$\begin{pmatrix} \delta & \nu'_{11} & \cdots & \nu'_{1n} \\ \delta & \nu_{11} & \cdots & \nu_{1n} \\ \cdot & \cdot & & \cdot \\ \cdot & \cdot & & \cdot \\ \cdot & \cdot & & \cdot \\ \delta & \nu_{n1} & \cdots & \nu_{nn} \end{pmatrix} \begin{pmatrix} -\delta \\ \mu_1 \\ \cdot \\ \cdot \\ \cdot \\ \mu_n \end{pmatrix} = \begin{pmatrix} 0 \\ \\ \\ \\ \\ 0 \end{pmatrix}.$$

It follows from Kramer's rule that the determinant of the coefficient matrix for (ii) must vanish. If we expand by minors of the first column we obtain

$$\rho - \rho' + \rho_3 - \cdots + (-1)^n \rho_{n+1} = 0,$$

where ρ_i, for $i \geqslant 3$, is the minor of the ith element of the first row.

Note that ρ_i is a sum of products of elements of the algebras $M(A_j)$ for $j \neq i - 1$. It follows that ρ_i is concentrated on $\Sigma_{j \neq i-1} A_j$ and, hence, restricts to zero on B. Since this is true for each $i = 3, \cdots, n + 1$, we conclude that $\rho - \rho'$ also restricts to zero on B.

The significance of the residue measure, for us, is the following. Under appropriate conditions on $(\mu_1, \cdots, \mu_n)$, a residue measure ρ for this tuple must have the property that $\rho|_B$ is nonzero and absolutely continuous. We shall prove this using Laplace transforms and the Cauchy-Weil integral formula. It provides us with a powerful tool for constructing absolutely continuous measures in certain L-subalgebras of $M(G)$.

7.2.4. Laplace Transforms. The subsemigroup A of G has the property that $\hat{A}$ is the set of functions of the form $(g, x) \to \gamma(g)e^{izx}$ for $\gamma \in \hat{G}_0$ and $z = (z_1, \cdots, z_n) \in \mathbf{C}^n$ with $\operatorname{Im}(z_j) \geq 0$ $(j = 1, \cdots, n)$. Similarly, for each A_i, $\hat{A}_i$ is the set of functions $(g, x) \to \gamma(g)e^{izx}$ for $\gamma \in \hat{G}_0$ and $z = (z_1, \cdots, z_n) \in \mathbf{C}^n$ with $\operatorname{Im}(z_j) \geq 0$ for $j \neq i$ and $\operatorname{Im}(z_i) = 0$.

We set

$$\Omega = G_0 \times \{z \in \mathbf{C}^n : \operatorname{Im}(z_j) \geq 0,\ j = 1, \cdots, n\},$$

and

$$\Omega_i = \hat{G}_0 \times \{z \in \mathbf{C}^n : \operatorname{Im}(z_j) \geq 0 \quad \text{for} \quad j \neq i,\ \operatorname{Im}(z_i) = 0\}.$$

If $\mu \in M(A)$ (or $M(A_i)$) we define the Laplace transform $\hat{\mu}$ by

$$\hat{\mu}(\gamma, z) = \int \gamma(g)e^{izx}\, d\mu(g, x) \quad \text{for } (\gamma, z) \in A \ \ (\text{or } A_i).$$

Note that since $\hat{G}_0$ is discrete, Ω is a discrete union of copies of the set $\{z \in \mathbf{C}^n : \operatorname{Im}(z_j) \geq 0, j = 1, \cdots, n\}$. The boundary of Ω is the union of "faces" Ω_i, while the interior $\operatorname{int}(\Omega)$ of Ω is a discrete union of copies of the domain $\{z \in \mathbf{C}^n : \operatorname{Im}(z_j) > 0, j = 1, \cdots, n\}$. If $\mu \in M(A)$ then $\hat{\mu}$ is holomorphic on $\operatorname{int}(\Omega)$, continuous on Ω, and bounded. If μ has compact support, then $\hat{\mu}$ is holomorphic on all of $\hat{G}_0 \times \mathbf{C}^n$.

The restriction of $\hat{\mu}$ to $\hat{G}_0 \times \{z \in \mathbf{C}^n : \operatorname{Im}(z_j) = 0, j = 1, \cdots, n\} = \hat{G}$ is (modulo a sign convention) the Fourier transform of μ. Note that $\hat{G} = \cap_i \Omega_i$.

The main theorem of this section is the following.

7.2.5. Theorem. *Suppose* $\mu_1, \cdots, \mu_n$ *are compactly supported measures in* $M(A)$ *such that*

(a) *the equation* $\mu_1 * \nu_{i1} + \cdots + \mu_n * \nu_{in} = \delta$ *has a solution in* $M(A_i)$ *for each* i;

(b) $\hat{\mu}$ *is bounded away from zero at infinity in* Ω;

(c) *the triple* $(\hat{\mu}_1, \cdots, \hat{\mu}_n)$ *has at least one common zero in* $\operatorname{int}(\Omega)$.

Then the residue measure $\rho = \det(\nu_{ij})$ has the property that $\rho|_B$ is nonzero and absolutely continuous.

This is a simplified version of a result in Taylor [4, Lemma 5.3] which eventually led to most of the results of Chapters 5 through 9. The original proof uses a vast amount of combinatorial machinery. The proof we shall outline here is much shorter, but only because we use the Cauchy-Weil integral formula to substitute for this combinatorial machinery.

It is worthwhile giving a separate proof for the case $n = 1$, since in this case the argument is quite transparent.

7.2.6. The Case $n = 1$. Here $G = G_0 \times R$, $A = G_0 \times R^+$, and the n-tuple $(\mu_1, \cdots, \mu_n)$ consists of a single measure $\mu = \mu_1 \in M(A)$. The equation in 7.2.5(a) becomes $\mu * \nu = \delta$. Hence, by hypothesis we have that μ has an inverse $\nu \in M(G)$. The residue measure ρ is simply ν.

Now Ω is the discrete union of half planes of the form $\{g_0\} \times \{z \in \mathbf{C} : \operatorname{Im} z > 0\}$

and, by (b) and (c) of 7.2.5, $\hat{\mu}$ is bounded away from zero at infinity on Ω and has a finite, nonempty set of zeros which occur in $\operatorname{int}(\Omega)$. It follows that the Fourier transform of ρ has a meromorphic extension, $\hat{\rho} = (\hat{\mu})^{-1}$, to all of Ω which is bounded at infinity.

Note that $B = G_0 \times \{x \in R : x < 0\}$. We define a functional F on $M(B)$ by

$$F(\lambda) = \frac{1}{2\pi} \sum_{\gamma \in \hat{G}_0} \int_{C_\gamma} (\lambda^1)^\wedge (\gamma, z)\hat{\rho}(\gamma, z)\, dz,$$

where $\lambda^1 \in M(A)$ is defined by $\lambda^1(E) = \lambda(-E)$ and, for each $\gamma \in \hat{G}_0$, C_λ is a Jordan curve in the upper half plane which encircles the poles of $\hat{\rho}$ which occur in $\{\gamma\} \times \{\operatorname{Im} z \geqslant 0\}$. Note that all but finitely many terms in the sum are zero since $\hat{\rho}$ has only finitely many poles.

The functional F is obviously continuous on $M(B)$ in the topology induced by the bounded continuous functions on B. Hence, $F(\lambda) = \int f\, d\lambda$ for some bounded continuous function f.

Now if λ is compactly supported on B and $\lambda = hm$ for some $h \in L^2(G) * L^2(G)$ (m is Haar measure), then $(\lambda^1)^\wedge$ is integrable on $\hat{G} = \hat{G}_0 \times R$ and decreases exponentially as $\operatorname{Im} z \to \infty$ on Ω. From the Cauchy integral theorem we conclude that

$$F(\lambda) = \frac{1}{2\pi} \sum_{\gamma} \int_R (\lambda^1)^\wedge (\gamma, x)\hat{\rho}(\gamma, x)\, dx.$$

This, by the Fourier inversion formula, yields

$$F(\lambda) = (\rho * h^1)(0) = \int h\, d\rho,$$

where $h^1(g_0, x) = h(-g_0, -x)$. Hence,

$$\int fh\, dm = \int h\, d\rho.$$

We conclude that the restriction of ρ to B is absolutely continuous with Radon-Nikodym derivative f.

If the measure $\rho|_B$ were zero, then the function F would necessarily be zero. Obviously this can happen only if $\hat{\rho}$ has no poles on Ω, i.e., only if $\hat{\mu}$ has no zeros on Ω. Since $\hat{\mu}$ has at least one zero, $\rho|_B$ is a nonzero absolutely continuous measure. This completes the proof of 7.2.5 in the case $n = 1$.

7.2.7. THE GENERAL CASE. The proof of 7.2.5 for general μ follows the same pattern. However, it is greatly complicated by combinatorial considerations. Without a lengthy analysis of the several variable Cauchy formula, which would take us far afield, we can only give a rough outline of the proof.

For each $\gamma \in \hat{G}_0$, let C_γ be a $2n$-cell in $\mathbf{C}^n_+ = \{z \in \mathbf{C}^n : \operatorname{Im}(z_i) \geqslant 0 \text{ for each } i\}$ with one vertex at the origin and edges parallel to the coordinate axes. Let C_γ be large enough that the zero set of $\hat{\mu}$ on $\{\gamma\} \times \mathbf{C}^n_+$ is contained in $\{\gamma\} \times (\operatorname{int} C_\gamma) \cup (\partial \mathbf{C}^n_+ \cap C_\gamma)$. Let $C^1_\gamma, \cdots, C^{4n}_\gamma$ be the $(2n-1)$-faces of C_γ. Note that n of these lie on the faces $\{z \in C^n_+ : \operatorname{Im}(z_i) = 0\}$ comprising $\partial \mathbf{C}^n_+$. We assume these are the faces $C^1_\gamma, \cdots, C^n_\gamma$.

We claim that for each face C^i_γ there is a solution to the equation

(i) $$\hat{\mu}_1(\gamma, z) f_{i1}(z) + \cdots + \hat{\mu}_n(\gamma, z) f_{in}(z) = 1$$

with each f_i continuous on C_γ^i and analytic in the $(n-1)$ complex variables that are unrestricted on C_γ^i. In fact, we may choose $f_{ij}(z) = \hat{\nu}_{ij}(\gamma, z)$ (where the ν_{ij} come from 7.2.5(a) for $i = 1, \cdots, n$ and $f_{i1}(z) = \hat{\mu}_1(\gamma, z)^{-1}$, $f_{ij} = 0$ $(j > 1)$ for $i > n$.

Now for a given tuple $\sigma = (i_1, \cdots, i_n)$ of n integers between 1 and $4n$ we set $C_\gamma^\sigma = C_\gamma^{i1} \cap \cdots \cap C_\gamma^{in}$. Note that C_γ^σ is an n-cell and the differential form

$$\det (f_{i_k j})_{kj} dz_{i1} \wedge \cdots \wedge dz_{in}$$

is defined on C_γ^σ and is independent of the ordering of the tuple $(i_1, \cdots, i_n)$ (since interchanging two indices changes the sign of both $\det(f_{i_k j})$ and $dz_{i_1} \wedge \cdots \wedge dz_{in}$).

Now it follows from the combinatorial machinery in the development of the several variable Cauchy formula (cf. Gleason [2]) that for an appropriate choice of orientation for each C_γ^σ and for ϕ continuous on $\mathbf{C}_+^n$ and analytic on int $\mathbf{C}_+^n$, the expression

(ii) $$F_\gamma(\phi) = \frac{1}{(2\pi)^n} \sum \int_{C_\gamma{}^\sigma} \phi(z) \det (f_{i_k j}(z))\, dz_{i_1} \wedge \cdots \wedge dz_{in}$$

is independent of the choice of the solution f_{ij} to (i) and the size of C_γ (as long as all common zeros of $\hat{\mu}_1, \cdots, \hat{\mu}_n$ are contained in int C_γ). The summation in (ii) is carried out over all *sets* $\sigma = \{i_1, \cdots, i_n\}$ of n integers between 1 and $4n$ without regard to ordering. We claim without proof that the functional F_γ is identically zero if and only if $\hat{\mu}_1(\gamma, z), \cdots, \hat{\mu}_n(\gamma, z)$ have no common zeros in int (C_γ). A proof of this fact follows from a careful analysis of the combinatorial arguments involved in the development of the several variable Cauchy theorem (Gleason [2]). The functional $\phi \to F_\gamma(\phi)$ turns out to be a finite linear combination of the power series coefficients of ϕ at the common zeros of $\hat{\mu}_1, \cdots, \hat{\mu}_n$.

As before, we define a functional F on compactly supported measures on B by

$$F(\lambda) = \sum_\gamma F_\gamma(\hat{\lambda}^1(\gamma, \cdot))$$

where $\lambda^1(g, x) = \lambda(-g, -x)$. This is continuous in the topology induced on measures by the bounded continuous functions on B and, therefore, has the form $F(\lambda) = \int f\, d\lambda$ for some bounded continuous function f on B. For λ compactly supported in B of the form $\lambda = hm$, with $h \in L^1(B)$ and $\hat{h} \in L^1(\hat{G})$, we have that $\hat{\lambda}^1(\gamma, z)$ is integrable on $\hat{G} = \hat{G}_0 \times R^n$ and vanishes exponentially as $\operatorname{Im}(z) \to \infty$.

Now the functions f_{ij} can be chosen from among the functions $\hat{\nu}_{ij}(\gamma, \cdot), \hat{\mu}(\gamma, \cdot)^{-1}$, and zero. Since $\hat{\mu}^{-1}$ is bounded at infinity on $\Omega = \hat{G}_0 \times \mathbf{C}_+^n$ and the $\hat{\nu}_{ij}$'s are bounded on faces of Ω, we may pass to the limit in (ii) as the cells C_γ become infinitely large and conclude that

$$F(\lambda) = \frac{1}{(2\pi)^n} \sum_\gamma \int_{R^n} \hat{\lambda}^1(\gamma, x) \det (\hat{\nu}_{ij}(\gamma, x))\, dx_1 \cdots dx_n$$

$$= \frac{1}{(2\pi)^n} \sum_\gamma \int_{R^n} \hat{\lambda}^1(\gamma, x) \hat{\rho}(\gamma, x)\, dx_1 \cdots dx_n$$

$$= h^1 * \rho(0) = \int h\, d\rho.$$

In other words, in the limit the contribution to (ii) from the C^σ_γ go to zero except for $\sigma = (1, \cdots, n)$.

As before, we conclude from the above that $\rho|_B$ is a nonzero measure which is absolutely continuous and has f as its Radon-Nikodym derivative.

This, in brief outline, is a proof of Theorem 7.2.5.

§7.3. Spectrum of a half-algebra

7.3.1. This section contains another technical fact that we shall need for our assault on Theorem 7.1.4. Here we shall assume that T is a locally compact topological semigroup with identity e, $\mathfrak{M}$ is an L-subalgebra of $M(T)$, and that the spectrum of $\mathfrak{M}$ consists of $\hat{T}$ with points identified according to the equivalence relation $f \sim g$ if $\hat{\mu}(f) = \int f \, d\mu = \int g \, d\mu = \hat{\mu}(g)$ for all $\mu \in \mathfrak{M}$.

We shall suppose that there is a continuous homomorphism $\alpha: T \to R$. We then cut T and $\mathfrak{M}$ in half by setting

$$T_+ = \{t \in T : \alpha(t) \geqslant 0\}, \quad T_- = \{t \in T : \alpha(t) \leqslant 0\},$$

and

$$\mathfrak{M}_+ = \mathfrak{M} \cap M(T_+), \quad \mathfrak{M}_- = \mathfrak{M} \cap M(T_-).$$

Our objective here is to compute the spectrum of $\mathfrak{M}_+$.

7.3.2. THE SEMICHARACTERS k_x. For each $x \in R$, the function k_x on T defined by $k_x(t) = e^{-x\alpha(t)}$ is a continuous (though not bounded) semicharacter of T. On T_+, the semicharacter k_x is bounded if $x \geqslant 0$ and, hence, is an element of $\hat{T}_+$.

We define the function k_∞ on T_+ by $k_\infty(t) = 0$ if $\alpha(t) > 0$ and $k_\infty(t) = 1$ if $\alpha(t) = 0$. Then k_∞ is a bounded Borel semicharacter on T_+ and k_x decreases monotonically to k_∞, pointwise, as $x \to \infty$. We clearly have

PROPOSITION. *For $x \in [0, \infty]$, $f \in \hat{T}$, the map $\mu \to \hat{\mu}(k_x f) = \int k_x f \, d\mu$ is a complex homomorphism of $\mathfrak{M}_+$.*

We shall show that every complex homomorphism of $\mathfrak{M}_+$ has this form. We do this by proving that for $\mu_1, \cdots, \mu_n \in \mathfrak{M}_+$ the equation $\mu_1 * \nu_1 + \cdots + \mu_n * \nu_n = \delta$ has a solution in $\mathfrak{M}_+$ if and only if the functions $\hat{\mu}_1, \cdots, \hat{\mu}_n$ do not vanish simultaneously at any $k_x f$. The proof of this fact involves piecing together solutions of linear equations in the algebras defined below:

7.3.3. THE ALGEBRAS $\mathfrak{M}_I$. We let $\mathfrak{M}_c$ denote the subalgebra of $\mathfrak{M}$ consisting of measures with compact support on T. For each $x \in R$ the map $\mu \to k_x \mu$ is an endomorphism of $\mathfrak{M}_c$ onto itself; that is, $\mu \to k_x \mu$ preserves convolution products. This follows from the fact that k_x is a continuous semicharacter on T.

On $\mathfrak{M}_c$ we define a norm $\| \ \|_x$ for each x by $\|\mu\|_x = \|k_x \mu\|$. For a closed interval $I = [x, y]$ and $\mu \in \mathfrak{M}_c$ we define $\|\mu\|_I = \max\{\|\mu\|_x, \|\mu\|_y\}$. Note that $\| \ \|_x$ satisfies $\|\mu * \nu\|_x = \|k_x(\mu * \nu\| \leqslant \|k_x \mu\| \, \|k_x \nu\| = \|\mu\|_x \, \|\nu\|_x$ and $\|\delta\|_x = \|k_x \delta\| = \|\delta\| = 1$ (since $k_x(e) = 1$ for the identity e of T). It follows that $\|\mu * \nu\|_I \leqslant \|\mu\|_I \, \|\nu\|_I$ and $\|\delta\|_I = 1$ for each interval I. Thus, the completion of $\mathfrak{M}_c$ in the norm $\| \ \|_I$ is a Banach algebra with identity for each I.

DEFINITION. If I is a finite closed interval (possibly degenerate), we denote by $\mathfrak{M}_I$ the completion of $\mathfrak{M}_c$ in the norm $\| \ \|_I$. If $I = [x, +\infty]$ (or $[-\infty, x[$) then we let $\mathfrak{M}_I$ be the completion of $\mathfrak{M}_c \cap \mathfrak{M}_\pm$ in the norm $\| \ \|_x$.

Note that, for a finite interval $I = [x, y]$, the algebra $\mathfrak{M}_I$ can be interpreted as the space of (possibly nonfinite) measures μ on T which satisfy the growth conditions implied by the statement $k_x \mu$ and $k_y \mu$ are finite measures. For an infinite interval, $I = [x, +\infty]$ or $I = [-\infty, x]$, the space $\mathfrak{M}_I$ is the space of measures in $\mathfrak{M}_x$ which are supported on $T_\pm$. In particular, $\mathfrak{M}_+ = \mathfrak{M}_{[0,\infty]}$ and $\mathfrak{M}_- = \mathfrak{M}_{[-\infty,0]}$.

7.3.4. PROPOSITION. (a) *For* $x < y$, $\mathfrak{M}_{[x,y]} = \mathfrak{M}_{[x,\infty]} + \mathfrak{M}_{[-\infty,y]}$;

(b) *if* $I \subset J$, *then* $\mathfrak{M}_J \subset \mathfrak{M}_I$;

(c) *if* $a < x < b$, *then* $\mathfrak{M}_{[a,x]} \cap \mathfrak{M}_{[x,b]} = \mathfrak{M}_{[a,b]}$;

(d) *if* $a \leqslant x \leqslant b$, *then the map* $\mu \to \hat{\mu}(k_x f) = \int k_x f \, d\mu$ *defines a complex homomorphism of* $\mathfrak{M}_{[a,b]}$ *for each* $f \in \hat{T}$.

PROOF. Each measure μ on T has a decomposition as $\mu = \nu + \omega$ with ν concentrated on T_+ and ω concentrated on T_-. If $x < y$ then $k_y \leqslant k_x$ on T_+, while $k_x \leqslant k_y$ on T_-. Hence, $\mu \in \mathfrak{M}_{[x,y]}$ if and only if $k_x \nu$ is finite and $k_y \omega$ is finite. Part (a) follows.

Since $x < y$ implies $k_y < k_x$ on T_+, it follows that $\mathfrak{M}_{[x,\infty]} \subset \mathfrak{M}_{[y,\infty]}$ for $x < y$. Similarly, $\mathfrak{M}_{[-\infty,y]} \subset \mathfrak{M}_{[-\infty,x]}$ for $x < y$. With this observation, part (b) follows directly from part (a).

It follows immediately from the definition that $\mathfrak{M}_{[a,x]} \cap \mathfrak{M}_{[x,b]} \subset \mathfrak{M}_{[a,b]}$. The reverse containment follows from (b). Part (d) is obvious.

7.3.5. PROPOSITION. *If* $\mu_1, \cdots, \mu_n \in \mathfrak{M}_{[a,b]}$, $x \in [a,b]$, *and the functions* $\hat{\mu}_1(k_x f), \cdots, \hat{\mu}_n(k_x f)$ *do not vanish simultaneously for any* $f \in \hat{T}$, *then the equation*

(i) $$\mu_1 * \nu_1 + \cdots + \mu_n * \nu_n = \delta$$

has a solution in $\mathfrak{M}_I$ *for some interval* $I \subset [a, b]$ *which contains* x *as an interior point in the relative topology of* $[a, b]$.

PROOF. We first assume that x is finite. By hypothesis, the elements $k_x \mu_1, \cdots, k_x \mu_n$ are in $\mathfrak{M}$ and have Gelfand transforms $(k_1 \mu_1)\,\hat{}, \cdots, (k_n \mu_n)\hat{}$ which do not vanish simultaneously on $\hat{T}$. Hence, there exist $\omega_1, \cdots, \omega_n \in \mathfrak{M}$ such that

(ii) $$(k_x \mu_1) * \omega_1 + \cdots + (k_x \mu_n) * \omega_n = \delta.$$

If we define $\rho_i = k_{-x} \omega_i$, then $\rho_i \in \mathfrak{M}_x$ and

(iii) $$\mu_1 * \rho_1 + \cdots + \mu_n * \rho_n = \delta.$$

If we replace each ρ_i by an appropriately close measure ρ_i' with compact support, we can achieve

(iv) $$\| \mu_1 * \rho_1' + \cdots + \mu_n * \rho_n' - \delta \|_x < 1.$$

Now $k_y\mu_i \in \mathfrak{M}$ and varies continuously with y for $y \in [a, b]$. Hence, inequality (iv) will continue to hold if we replace each μ_i and ρ_i' by $k_c\mu_i$ and $k_c\rho_i'$ provided c is sufficiently small and $x + c \in [a, b]$. It follows that we can choose an interval $I \subset [a, b]$ with x as a relative interior point and

$$\| \mu_1 * \rho_1' + \cdots + \mu_n * \rho_n' - \delta \|_I < 1.$$

We conclude that $\mu_1 * \rho_1' + \cdots + \mu_n * \rho_n'$ has an inverse λ in $\mathfrak{M}_I$. Thus $\nu_i = \lambda * \rho_i'$ $(i = 1, \cdots, n)$ yields the required solution of (i). This completes the proof in the case where x is finite.

If $x = b = +\infty$, the measures $\mu_1, \cdots, \mu_n$ are supported on T_+ and the measures $k_x\mu_1, \cdots, k_x\mu_n$ are elements of $\mathfrak{M}^0 = \mathfrak{M} \cap M(T_0)$, where $T_0 = T_+ \cap T_-$ (recall that k_∞ is defined on T_+ by $k_\infty(t) = 1$ for $t \in T_0$ and $k_\infty(t) = 0$ for $t \in T_+ \backslash T_0$). By hypothesis, the functions $(k_x\mu_1)^\wedge, \cdots, (k_x\mu_n)^\wedge$ do not vanish simultaneously on $\hat{T}$. Now, by Lemma 4.5.6, each complex homomorphism of $\mathfrak{M}^0$ extends to a complex homomorphism of $\mathfrak{M}$ and, hence, is determined by a point of $\hat{T}$. Thus the functions $(k_x\mu_1)^\wedge, \cdots, (k_x\mu_n)^\wedge$ do not vanish simultaneously on the spectrum of $\mathfrak{M}^0$.

We conclude from the above that the equation $(k_x\mu_1) * \omega_1 + \cdots + (k_x\mu_n) * \omega_n = \delta$ has a solution $\omega_1, \cdots, \omega_n \in \mathfrak{M}^0$. Since each $k_y\mu_i$ varies continuously as an element of $\mathfrak{M}_+$ for $y \in [a, \infty]$, we may proceed as before, concluding that

$$\| \mu_1 * \omega_1 + \cdots + \mu_n * \omega_n - \delta \|_I < 1$$

for $I = [y, \infty]$, with $a \leqslant y \leqslant \infty$ and y sufficiently large. By inverting $\mu_1 * \omega_1 + \cdots + \mu_n * \omega_n$ in $\mathfrak{M}_I$ and multiplying the result by each ω_i, we obtain the desired solution $\nu_1, \cdots, \nu_n$.

7.3.6. Proposition. *If* $\mu_1, \cdots, \mu_n \in \mathfrak{M}_{[a,b]}$, $a < x < b$, *and the equation*

(i) $$\mu_1 * \nu_1 + \cdots + \mu_n * \nu_n = \delta$$

has a solution in $\mathfrak{M}_{[a,x]}$ *and a solution in* $\mathfrak{M}_{[x,b]}$, *then it also has a solution in* $\mathfrak{M}_{[a,b]}$.

Proof. Let $\rho_1, \cdots, \rho_n \in \mathfrak{M}_{[a,x]}$ and $\omega_1, \cdots, \omega_n \in \mathfrak{M}_{[x,b]}$ be solutions of (i). For each i, j we set

$$\lambda_{ij} = \det \begin{pmatrix} \rho_i & \rho_j \\ \omega_i & \omega_j \end{pmatrix} = \rho_i * \omega_j - \rho_j * \omega_i \in \mathfrak{M}_x.$$

Note that $\Sigma_i \mu_i * \lambda_{ij} = \omega_j - \rho_j$ for each j.

By 7.3.4 we may choose $\lambda_{ij}' \in \mathfrak{M}_{[-\infty,x]} \subset \mathfrak{M}_{[a,x]}$ and $\lambda_{ij}'' \in \mathfrak{M}_{[x,\infty]} \subset \mathfrak{M}_{[x,b]}$ such that

$$\lambda_{ij} = \lambda_{ij}' - \lambda_{ij}'' \quad \text{and} \quad \lambda_{ij}' = -\lambda_{ji}', \quad \lambda_{ij}'' = -\lambda_{ji}''.$$

Hence $\nu_j = \omega_j - \Sigma_i \mu_i * \lambda_{ij}' = \rho_j - \Sigma_i \mu_i * \lambda_{ij}''$ is an element of $\mathfrak{M}_{[a,x]} \cap \mathfrak{M}_{[x,b]} = \mathfrak{M}_{[a,b]}$ for each j and

$$\sum_j \mu_j * \nu_j = \sum_j \mu_j * \omega_j - \sum_{i,j} \mu_j * \mu_i * \lambda_{ij}'.$$

Since $\lambda'_{ij} = -\lambda'_{ij}$, we conclude that

$$\sum_j \mu_j * \nu_j = \sum_j \mu_j * \omega_j = \delta.$$

Hence $\nu_1, \cdots, \nu_n \in \mathfrak{M}_{[a,b]}$ is the required solution of (i).

7.3.7. PROPOSITION. *If* $\mu_1, \cdots, \mu_n \in \mathfrak{M}_+ = \mathfrak{M}_{[0,\infty]}$ *and* $\hat{\mu}_1(k_x f), \cdots, \hat{\mu}_n(k_x f)$ *do not vanish simultaneously for* $x \in [0, \infty], f \in \hat{T}$, *then the equation*

(i) $$\mu_1 * \nu_1 + \cdots + \mu_n * \nu_n = \delta$$

has a solution for $\nu_1, \cdots, \nu_n \in \mathfrak{M}_+$.

PROOF. By 7.3.5 there exists a solution to (i) in $\mathfrak{M}_{[x,\infty]}$ for sufficiently large x. If $x_0 = \inf \{x:$ (i) has a solution in $\mathfrak{M}_{[x,\infty]}\}$, then 7.3.5 also implies that (i) has a solution in $\mathfrak{M}_{[a,x]}$ for some interval $[a, x] \subset [0, \infty]$ with x_0 as a relative interior point; that is, $x_0 < x$ and either $0 = a = x_0$ or $0 \leqslant a < x_0$. By 7.3.6 the solutions in $\mathfrak{M}_{[a,x]}$ and $\mathfrak{M}_{[x,\infty]}$ may be used to construct a solution in $\mathfrak{M}_{[a,\infty]}$. By the choice of x_0, we conclude that $x_0 = 0$ and (i) has a solution in $\mathfrak{M}_{[0,\infty]} = \mathfrak{M}_+$.

§7.4. An alternate representation of $\mathfrak{M}$

7.4.1. In this section we assume that $\mathfrak{M}$ is a commutative semisimple measure algebra with norm one identity δ and structure semigroup S. We also assume that $1 \in \hat{S}$ is a critical point. We let Γ be the locally compact group $\{f \in \hat{S}: |f| = 1\}$; i.e., Γ is the maximal group at the identity in $\hat{S}$. We let G be an l.c.a. group with Γ as dual group and $K =$ kernel (S) as Bohr compactification.

The object of this section is to represent $\mathfrak{M}$ on a new semigroup T which is only locally compact, but which has G as its kernel (minimal ideal). The semigroup T will be a subsemigroup of S with a stronger topology.

7.4.2. THE SEMIGROUP T. The cartesian product $G \times S$ is a locally compact topological semigroup if the operation is defined coordinatewise. There are two continuous homomorphisms of $G \times S$ into K. One of these sends $(g, s) \in G \times S$ to $\alpha(g) \in K$, where $\alpha: G \to K$ is the natural embedding of G in its Bohr compactification. The other sends (g, s) to $ps \in K$, where p is the idempotent in K. The subset on which these maps agree is clearly a closed subsemigroup of $G \times S$. This is the subsemigroup $T_1 = \{(g, s) \in G \times S: \alpha(g) = ps\}$.

The projection $(g, s) \to s: G \times S \to S$ is injective on T_1. In fact, if $(g_1, s) \in T$ and $(g_2, s) \in T_1$, then $\alpha(g_1) = ps = \alpha(g_2)$; this implies $g_1 = g_2$ since α is injective. We let T, as a semigroup, be the projection of T_1 in S. However, we give T the topology which makes this projection a homeomorphism. The topology of T can be described by specifying which sets are compact (since T is locally compact).

PROPOSITION. *As a subsemigroup of* S, T *is* $\{s \in S: ps \in \alpha(G)\}$. *In the topology of* T, *a set* E *is compact if and only if* $pE \subset K$ *is the image under* α *of a compact subset of* G.

PROOF. The first contention is apparent from the definition of T.

A subset $E \subset T$ is compact if and only if $E_1 = \{(g, s) \in T_1 : s \in E\}$ is compact. However, E_1 is compact if and only if its projection E' on G is compact. Since $\alpha(E') = pE$, the proof is complete.

7.4.3. PROPOSITION. *The map $\alpha : G \to K \subset S$ maps G onto the minimal ideal of T and is a homeomorphism if this ideal is given the topology of T.*

PROOF. If $g \in G$ and $s = \alpha(g) \in K$, then $(g, s) \in T_1$ and hence $s \in T$. Thus α maps G into T. The image of α is a group and thus to prove that it is the minimal ideal of T, it suffices to prove that it is an ideal of T. However, if $s = \alpha(g)$ and $t \in T$ with $pt = \alpha(g')$, then $\alpha(gg') = spt = st$; hence $\alpha(G)$ is an ideal. That α is a homeomorphism onto its image is apparent from the description of the topology of T in 7.2.2.

7.4.4. PROPOSITION. *Each measure in $\mathfrak{M}$ is concentrated on T and is a regular Borel measure relative to the T topology.*

PROOF. Our description in 7.2.2 of the compact subsets of T makes it apparent that $\mu \in M(S)$ will be concentrated on T and regular for the T topology if and only if the measure μ' on K, defined by $\mu'(E) = \mu(\{s \in S : ps \in E\})$, is $\nu \circ \alpha^{-1}$ for some $\nu \in M(G)$. Note that $\mu' = \delta_p * \mu$, where $\delta_p \in M(S)$ is the point mass at p.

Now the image of $M(G)$ in $M(K)$ under $\nu \to \nu \circ \alpha^{-1}$ consists of those measures on K whose Fourier transforms are continuous on Γ (Rudin [5]). Since $(\delta_p * \mu)^\wedge(f) = f(p)\hat{\mu}(f) = \hat{\mu}(f)$ for $f \in \Gamma, \mu \in \mathfrak{M}$, we conclude that $\delta_p * \mu = \nu \circ \alpha^{-1}$ for some $\nu \in M(G)$. The proposition follows.

If we identify G with the kernel of T, as 7.2.3 allows us to do, then we have the following

7.4.5. PROPOSITION. *If 1 is a critical point for $\mathfrak{M}$, then there is an abelian, locally compact topological semigroup T and a representation of $\mathfrak{M}$ as an L-subalgebra of $M(T)$ such that*

(a) *every complex homomorphism of $\mathfrak{M}$ is determined by an element of $\hat{T}$;*

(b) *the kernel of T is an l. c. a. group G;*

(c) *if $\hat{T}$ is given the Gelfand topology induced by $\mathfrak{M}$, then the dual group Γ of G is homeomorphically embedded as an open subgroup $\{f \in \hat{T} : |f| = 1\}$ of $\hat{T}$.*

PROOF. The proposition follows if we simply observe that each element of $\hat{S}$ (the spectrum of $\mathfrak{M}$) restricts to an element of $\hat{T}$.

We should point out one minor problem here. Although $\mathfrak{M}$ was weak-* dense in $M(S)$, it may fail to be weak-* dense in $M(T)$. Thus, $\mathfrak{M}$ may fail to separate points in $\hat{T}$. If we wish to interpret $\hat{T}$ as the spectrum of $\mathfrak{M}$, we must introduce the equivalence relation $f \sim g$ if and only if $\int f\, d\mu = \int g\, d\mu$ for every $\mu \in \mathfrak{M}$. Note that, for this relation, each equivalence class contains exactly one element which is the restriction of an element of $\hat{S}$.

7.4.6. A MAP OF $\mathfrak{M}$ INTO $M(G)$. If $p \in G \subset T$ is the idempotent of G and $\delta_p \in M(T)$ is the point mass at p, then $\delta_p * \mu \in M(G)$ for each $\mu \in \mathfrak{M}$. In fact, the map $\mu \to \delta_p * \mu : \mathfrak{M} \to M(G)$ is just the map of measures induced by the continuous homomorphism $t \to pt : T \to G$. It follows that $\mu \to \delta_p * \mu$ is a homomorphism of measure algebras. We would like to know that the image $\delta_p * \mathfrak{M}$ of $\mathfrak{M}$ in $M(G)$ is weak-* dense.

Now the fact that $\delta_p * \mathfrak{M}$ is an L-subalgebra of $M(G)$ implies that its weak-* closure has the form $M(V)$, where V is a closed subsemigroup of G. Note that V cannot be contained in any proper closed subgroup of G. If it were, then $\mathfrak{M}$ would not separate points of Γ which agreed on this subgroup.

If V were proper in G, then 4.5.2 would imply that 1 was not isolated in $\hat{V}^+$ in the compact-open topology. Since elements of $\hat{V}$ determine complex homomorphisms of $\delta_p * \mathfrak{M} \subset M(V)$ and hence of $\mathfrak{M}$, this would violate the assumption that 1 is a critical point of the spectrum of $\mathfrak{M}$. Hence $V = G$ and

PROPOSITION. *The space* $\delta_p * \mathfrak{M}$ *is weak-* dense in* $M(G)$.

§7.5. **Proof of Theorem** 7.1.4.

7.5.1. We assume that $\mathfrak{M}$ is a semisimple commutative measure algebra with norm one identity δ and that 1 is a critical point for $\mathfrak{M}$. Let T, $G = \text{kernel}\ (T)$, and $\Gamma = \{f \in \hat{T} : |f| = 1\} \simeq \hat{G}$ be as in §7.4. In order to prove Theorem 7.1.4, we must show that $\mathfrak{M} \cap M(G)$ is a subalgebra of $\mathfrak{M}$ which contains $L(G)$ and is contained in Rad $(L(G))$. It turns out that everything is easy except proving that $\mathfrak{M} \cap L(G) \neq 0$. We attack this part of the problem first.

Initially, we shall assume that the group G has the form $G_0 \times R^n$ for $n \geqslant 0$ and G_0 a compact subgroup of G.

7.5.2. NOTATION. As in §7.2, we let $A = G_0 \times \{x \in R^n : x_i \geqslant 0 \text{ for } i = 1, \cdots, n\}$, $A_j = G_0 \times \{x \in R^n : x_i \geqslant 0 \text{ for } i \neq j\}$, and $B = G_0 \times \{x \in R^n : x_i < 0,\ i = 1, \cdots, n\}$.

We let p be the idempotent in $G = \text{kernel}\ (T)$ and define a map $\alpha : T \to G$ by $\alpha(t) = pt$. Thus, α is a continuous homomorphism of T onto $G = G_0 \times R^n$.

For each $j = 1, \cdots, n$ we set $T_j = \alpha^{-1}(A_j)$ and $\mathfrak{M}_j = \mathfrak{M} \cap M(T_j)$. Similarly, $T_0 = \alpha^{-1}(A)$ and $\mathfrak{M}_0 = \mathfrak{M} \cap M(T_0)$. Each T_j is a closed subsemigroup of T containing the identity, and each $\mathfrak{M}_j$ is a closed L-subalgebra of $\mathfrak{M}$ containing δ.

7.5.3. THE SPECTRUM OF $\mathfrak{M}_j$. For each j, the algebra $\mathfrak{M}_j$ is obtained from $\mathfrak{M}$ by a repeated application of the "half-algebra" construction of §7.3. It follows from a repeated application of 7.3.7 that the spectrum of $\mathfrak{M}_j$ consists of complex homomorphisms of the form

$$\mu \longrightarrow \hat{\mu}(k_x f) = \int k_x f\, d\mu,$$

where $f \in \hat{T}$ and $k_x(t) = e^{-x \cdot \alpha(t)}$ for $x \in Q_j = \{x \in (R^+ \cup \{\infty\})^n : x_j = 0\}$. Thus,

PROPOSITION. *If* $\mu_1, \cdots, \mu_n \in \mathfrak{M}_j$ *and the functions* $\hat{\mu}_1(k_x f), \cdots, \hat{\mu}_n(k_x f)$ *do not vanish simultaneously for* $x \in Q_j, f \in \hat{T}$, *then the equation*

$$\mu_1 * \nu_1 + \cdots + \mu_n * \nu_n = \delta$$

has a solution $\nu_1, \cdots, \nu_n \in \mathfrak{M}_j$.

7.5.4. A SPECIAL MEASURE. We now invoke the hypothesis that 1 is a critical point. This means that 1 is an isolated point of $\hat{T}^+/(\sim) = \hat{S}^+$ in the strong topology. Thus, there exists a positive measure $\nu \in \mathfrak{M}$ and an $\epsilon > 0$ such that $\int |1 - f|\, d\nu < \epsilon$ for $f \in \hat{T}^+$ implies $f \sim 1$. We then have

$$\int f\, d\nu \leqslant \int 1\, d\nu - \epsilon$$

for $f \in \hat{T}^+$ with $f \not\sim 1$. If we normalize ν, then for an appropriate $r < 1$ this becomes

$$\hat{\nu}(f) = \int f\, d\nu = \|f\nu\| \leqslant r < 1$$

for all $f \in \hat{T}^+$ with $f \not\sim 1$.

Now it follows from 7.4.6 that the image of the map $\mu \to \mu \circ \alpha^{-1} : \mathfrak{M} \to M(G)$ is weak-$*$ dense in $M(G)$. Since $\mathfrak{M}_0 \doteq \mathfrak{M} \cap M(T_0)$ is just the inverse image of $M(A)$ under this map, there must be a normalized positive measure $\omega \in \mathfrak{M}_0$ such that $\omega \circ \alpha^{-1}$ is compactly supported in $\operatorname{int} A$. Now for a sufficiently large power ω^k of ω, we must have $\omega^k * \nu$ is nearly concentrated on α^{-1} $(\operatorname{int} A) \subset T_0$. That is, if we set $\mu = \omega^k * \nu|_{T_0}$, then for a given $\epsilon > 0$ we may choose k large enough that $\|\mu - \omega^k * \nu\| < \epsilon$. Now,

$$(\omega^k * \nu)\,\hat{}\,(f) = \hat{\omega}^k(f)\hat{\nu}(f) \leqslant r < 1 \quad \text{for} \quad f \in \hat{T}^+, f \not\sim 1.$$

It follows that we may choose a positive normalized measure $\mu \in \mathfrak{M}_0$, with $\mu \circ \alpha^{-1} \in M(\operatorname{int} A)$ and $\hat{\mu}(f) \leqslant r' < 1$ for $f \in \hat{T}^+$ with $f \not\sim 1$ and for some $r' < 1$.

Now if $x \in Q = (R^+ \cup \{\infty\})^n$, then the function k_x is bounded by 1 on T_0 and

$$|\hat{\mu}(k_x f)| = \left|\int k_x f\, d\mu\right| \leqslant \int |f|\, d\mu = \hat{\mu}(|f|) < r'$$

for each $f \in \hat{T}^+$ with $|f| \not\sim 1$, i.e., for every $f \in \hat{T}$ which is not equivalent to an element of Γ. Also, observe that $\hat{\mu}(k_x f) = 0$ for arbitrary $f \in \hat{T}$ if any coordinate of x is infinite.

If we choose r_1 with $r' < r_1 < 1$ and set $\mu_1 = r_1\delta - \mu$, then we conclude:

PROPOSITION. *The measure* $\mu_1 \in \mathfrak{M}_0$ *has the property that* $\hat{\mu}_1(k_x f) \neq 0$ *for* $x \in Q, f \in \hat{T}$ *unless* f *is equivalent to an element of* Γ *and* x *is finite. However, there is a finite* $x \in Q$ *for which* $\hat{\mu}_1(k_x) = 0$.

The fact that $\hat{\mu}_1(k_x) = 0$ for some x follows from the fact that $\mu \circ \alpha^{-1}$ is concentrated on $\operatorname{int}(A)$ and $\|\mu_1\| = 1$. This implies that as x increases, $\hat{\mu}(k_x)$ decreases continuously from 1 (at $x = 0$) and approaches 0 as $x \to \infty$. Hence, $\hat{\mu}(k_x) = r_1$ for some value of x.

7.5.5. AN n-TUPLE IN $\mathfrak{M}_0$. We wish to construct an n-tuple of measures $\mu_1, \cdots, \mu_n \in \mathfrak{M}_0$ such that $\hat{\mu}_1, \cdots, \hat{\mu}_n$ fail to vanish simultaneously on the spectrum of $\mathfrak{M}_j$ for $j = 1, \cdots, n$, but do vanish simultaneously at some point of the spectrum of $\mathfrak{M}_0$.

Now the spectrum of $\mathfrak{M}_0$ is determined by the elements $k_x f$ for $x \in Q, f \in \hat{T}$, while the spectrum of $\mathfrak{M}_j$ is determined by the points of this form with $x \in Q_j$.

We already have a measure μ_1 such that $\hat{\mu}_1$ does not vanish except on certain points $k_x f$ with $f \in \Gamma$. If we represent G as $G_0 \times R^n$ then each $f \in \Gamma|_G$ has the form $f(g_0, u) = \gamma(g_0)e^{iu\cdot y}$ for some $\gamma \in \hat{G}_0$ and $y \in R^n$. Thus the elements $k_x f$ for $f \in \Gamma$ and finite x have, on G, the form

$$(g_0, u) \longrightarrow \gamma(g_0)e^{iz\cdot u}$$

with $(\gamma, z) \in \hat{G}_0 \times \{z \in \mathbf{C}^n : \operatorname{Im}(z_i) \geqslant 0 \text{ for } i = 1, \cdots, n\} = \hat{G}_0 \times \Omega$.

Now for $x \in Q, f \in \Gamma, \mu \in \mathfrak{M}$ we have $(\delta_p * \mu)^\wedge (k_x f) = f(p)\hat{\mu}(k_x f) = \hat{\mu}(k_x f)$. Thus, when restricted to $\{k_x f : f \in \Gamma,\ x \text{ finite}\}$, $\hat{\mu}$ may be represented as a function $(\gamma, z) \to \hat{\mu}(\gamma, z)$ on $\hat{G}_0 \times \Omega$, where

$$\hat{\mu}(\gamma, z) = (\delta p * \mu)^\wedge (\gamma, z) = \int \gamma(g_0)e^{iz\cdot u}\, d(\delta p * \mu)(g_0, u).$$

Now $\hat{\mu}_1(\gamma, z) = 0$ for $\gamma = 1$ and some $z = ix$ $(x > 0)$, but $\hat{\mu}_1$ is bounded away from zero at infinity. Let $x_1, \cdots, x_n$ be the coordinates of x. The functions $(\gamma, z) \to (z_i - ix_i)(z_i + ix_i)^{-1}$ for $i = 2, \cdots, n$ have a common zero set on $G_0 \times \mathbf{C}^n$ consisting of the set $\Lambda = \{(\gamma, z) : z_i = ix_i \text{ for } i = 2, \cdots, n\}$. This set meets the zero set of $\hat{\mu}_1$ in a nonempty compact set $K \subset \hat{G}_0 \times \operatorname{int}(\Omega)$. Since G_0 is compact, $\hat{G}_0$ is discrete and K meets only finitely many of the sets $\{\gamma\} \times \operatorname{int}(\Omega)$ for $\gamma \in \hat{G}_0$. Now the intersection of Λ with each $\{\gamma\} \times \operatorname{int}(\Omega)$ is a copy of the complex half plane and $\hat{\mu}$ is analytic there. It follows that K must be a finite set.

The functions $(\gamma, z) \to (z_i - ix_i)(z_i + ix_i)^{-1}$ are transforms of measures $\nu_2, \cdots, \nu_n$ on A. Hence by 7.4.6 we may approximate $\nu_2, \cdots \nu_n$ by measures $\delta_p * \mu_2, \cdots, \delta_p * \mu_n$, with $\mu_i \in \mathfrak{M}_0$, in such a way that the common zero set for $\hat{\mu}_1, \cdots, \hat{\mu}_n$ on $\hat{G}_0 \times \Omega$ remains a compact subset of $\hat{G}_0 \times \operatorname{int}(\Omega)$ containing $(1, ix)$. However, since the functions $\hat{\mu}_1, \cdots, \hat{\mu}_n$ are analytic on the analytic manifold $\hat{G}_0 \times \operatorname{int}(\Omega)$, this common zero set must be finite (Gunning and Rossi [1]). Thus

PROPOSITION. *There exist measures* $\mu_1, \cdots, \mu_n \in \mathfrak{M}_0$ *such that*

(a) $\hat{\mu}_1(k_x f) \neq 0$ *for* $f \in \hat{T}^+$ *with* $|f| \not\to 1$ *or for* x *nonfinite*;

(b) *the common zero set of* $\hat{\mu}_1(k_x f), \cdots, \hat{\mu}_n(k_x f)$ *is a finite subset of* $\{k_x f : x \in \operatorname{int}(Q), f \in \Gamma\}$ *and contains at least one point of the form* k_x.

7.5.6. PROPOSITION. *Under the hypothesis of this section, and the additional hypothesis that* $G = G_0 \times R^n$ *with* G_0 *compact, we have* $\mathfrak{M} \cap L(G) \neq (0)$.

PROOF. If $n \geqslant 1$ let $\mu_1, \cdots, \mu_n \in \mathfrak{M}_0$ be the measures constructed above. By 7.5.3 we may choose, for each $j = 1, \cdots, n$, elements $\nu_{j1}, \cdots, \nu_{jn} \in \mathfrak{M}_j$ such that

$$\mu_1 * \nu_{j1} + \cdots + \mu_n * \nu_{jn} = \delta.$$

We then let $\rho = \det(\nu_{ij}) \in \mathfrak{M}$.

Now the homomorphism $\mu \to \mu \circ \alpha^{-1} : \mathfrak{M} \to M(G)$ maps $\mu_1, \cdots, \mu_n$ into $M(A)$ and $\nu_{j1}, \cdots, \nu_{jn}$ into $M(A_j)$. Thus, $\rho \circ \alpha^{-1}$ will be a residue measure for

$\mu_1 \circ \alpha^{-1}, \cdots, \mu_n \circ \alpha^{-1}$. It will be nonvanishing on B since $\hat{\mu}_i(k_x) = (\mu_i \circ \alpha^{-1})\hat{\ }(1, x) = 0$ for $i = 1, \cdots, n$ and for some $x \in \operatorname{int}(Q)$ (Proposition 7.2.5). Also, by 7.2.5, $(\rho \circ \alpha^{-1})|_B$ is absolutely continuous.

Now if we replace μ_i by $f\mu_i$ for $i = 1, \cdots, n, f \in \hat{T}$, and ν_{ij} by $f\nu_{ij}$ for $i, j = 1, \cdots, n$ then we have

$$(f\mu_1) * (f\nu_{j1}) + \cdots + (f\mu_n) * (f\nu_{jn}) = f\delta = \delta;$$

and so $f\rho = \det(f\nu_{ij})$ has the property that $(f\rho) \circ \alpha^{-1}$ is a residue measure in $M(G)$ for $(f\mu_1 \circ \alpha^{-1}, \cdots, f\mu_n \circ \alpha^{-1})$.

If $f \in \hat{T}^+$ and $f \not\sim 1$, then the functions $(f\mu_i)\hat{\ }(k_x g) = \hat{\mu}_i(k_x fg)$ do not vanish simultaneously for any $x \in Q, g \in \hat{T}$. It follows, in this case, that the equation

$$(f\mu_1) * \nu_1 + \cdots + (f\mu_n) * \nu_n = \delta$$

has a solution in $\mathfrak{M}_0$ and hence that the residue measure $(f\rho) \circ \alpha^{-1}$ restricts to zero on B. This implies that $f\rho$ restricts to zero on $T_0^- = \{t \in T : \alpha(t) \in B\}$ although ρ does not. Thus, if $\rho_1 = \rho|_{T_0^-}$, then $\rho_1 \neq (0), f\rho_1 = 0$ for each $f \in \hat{T}^+$ with $f \not\sim 1$, and $\rho_1 \circ \alpha^{-1}$ is absolutely continuous on G.

Now $T \backslash G$ is the union of the open sets $\{t \in T : f(t) > 0\}$ as f ranges over elements of $\hat{T}^+$ not equivalent to 1. This follows from the fact that $S \backslash K$ is the union of the sets $\{s \in S : f(s) > 0\}$ for $f \in \hat{S}^+$, $f < 1$, which in turn follows from the fact that $\hat{S}$ separates points in S. We conclude that the measure ρ_1 above is concentrated on G and that $\rho_1 = \rho_1 \circ \alpha^{-1}$ is absolutely continuous. Hence $\mathfrak{M} \cap L(G) \neq (0)$ and the proof is complete in the case $n \geqslant 1$.

If $n = 0$, then $G = G_0$ is compact and $\Gamma = \hat{G}_0$ is discrete. Since Γ is open in $\hat{T}/(\sim)$, we conclude that 1 is an isolated point of the spectrum of $\mathfrak{M}$. By the Shilov idempotent theorem, there is a measure $\mu \in \mathfrak{M}$ such that $\hat{\mu}(1) = 1$ and $\hat{\mu}(f) = 0$ for $f \in \hat{T}$ with $f \not\sim 1$. The only measure on T that can have this property is the Haar measure of G. Hence, $\mathfrak{M} \cap L(G) \neq (0)$ in this case as well.

7.5.7. THE GENERAL CASE. We now return to the case where G is arbitrary. By the structure theory for l. c. a. groups (Rudin [5]), G has an open subgroup G_1 of the form $G_0 \times R^n$ with G_0 compact. We let $T_1 = \{t \in T : pt \in G_1\}$ and $\mathfrak{M}_1 = \mathfrak{M} \cap M(T_1) = \{\mu \in \mathfrak{M} : \delta_p * \mu \in M(G_1)\}$.

Since G_1 is open in G, it follows from 7.4.6 that $\mathfrak{M}_1 \neq (0)$ and, in fact, $\delta_p * \mathfrak{M}_1$ is weak-$*$ dense in $M(G_1)$.

We claim that 1 is a critical point for the algebra $\mathfrak{M}_1$. Suppose the contrary. It follows from 4.5.6 that each complex homomorphism of $\mathfrak{M}_1$ extends to a complex homomorphism of $\mathfrak{M}$. Hence, there exists a net $\{f_\alpha\} \subset \hat{T}$ such that $0 \leqslant f_\alpha | T_1 < 1$ for each α and $\hat{\omega}(f_\alpha) \to \hat{\omega}(1)$ for each $\omega \in \mathfrak{M}_1$. We may assume for each α that $f_\alpha \in \hat{T}^+$ (otherwise replace f_α by $|f_\alpha|$). Now suppose that μ is a positive measure in $\mathfrak{M}$ concentrated on $T_g = \{t \in T : pt \in gG_1\}$ for some coset gG_1. Since $\delta_p * \mathfrak{M}$ is weak-$*$ dense in $M(G)$, there exists a positive $\nu \in \mathfrak{M}$ such that $\delta_p * \nu \in M(g^{-1}G_1)$, i.e., ν is concentrated on T_{g-1}. Then $\mu * \nu \in \mathfrak{M}_1$ and hence

$$\hat{\mu}(f_\alpha)\hat{\nu}(f_\alpha) = (\mu * \nu)^\wedge(f_\alpha) \longrightarrow (\mu * \nu)^\wedge(1) = \hat{\mu}(1)\hat{\nu}(1).$$

Since $\hat{\mu}(f_\alpha) \leqslant \hat{\mu}(1)$ and $\hat{\nu}(f_\alpha) \leqslant \hat{\nu}(1)$ for each α, we conclude that $\hat{\mu}(f_\alpha) \to \hat{\mu}(f)$. Since each measure in $\mathfrak{M}$ is the sum of an absolutely convergent series of measures concentrated on sets T_g, it follows that $\hat{\mu}(f_\alpha) \to \hat{\mu}(1)$ for every $\mu \in \mathfrak{M}$. This is impossible since 1 is a critical point for $\mathfrak{M}$ and $f_\alpha < 1$ for each α. The resulting contradiction establishes our claim that 1 is a critical point for $\mathfrak{M}_1$.

It now follows from 7.5.6 that $\mathfrak{M}_1 \cap L(G_1) \neq (0)$. However, $L(G_1) \subset L(G)$ and so we conclude

PROPOSITION. *Under the hypotheses of this section,* $\mathfrak{M} \cap L(G) \neq (0)$.

7.5.8. COMPLETION OF THE PROOF OF 7.1.4. We now have that $\mathfrak{M} \cap L(G) \neq (0)$. It is also an ideal of $\mathfrak{M}$. In fact, $\mathfrak{M} \cap M(G)$ is an ideal of $\mathfrak{M}$ since G is an ideal of T; furthermore, $L(G)$ is an ideal of $M(G)$. Thus, for $\nu \in \mathfrak{M} \cap L(G), \mu \in \mathfrak{M}$, we have $\mu * \delta_p \in M(G), \delta_p * \nu = \nu$, and $\mu * \nu = \mu * \delta_p * \nu \in \mathfrak{M} \cap L(G)$. Note that the same argument shows that $\nu * \mathfrak{M} = \nu * (\delta_p * \mathfrak{M})$for $\nu \in L(G) \cap \mathfrak{M}$. Since $\delta_p * \mathfrak{M}$ is weak-* dense in $M(G)$ we conclude that $\nu * M(G) \subset L(G) \cap \mathfrak{M}$ for $\nu \in L(G) \cap \mathfrak{M}$. Thus $L(G) \cap \mathfrak{M}$ is an L-ideal of $M(G)$. Since $L(G)$ contains no proper L-ideals, we conclude that $L(G) \subset \mathfrak{M}$.

Now the ideal $\mathfrak{N} = \mathfrak{M} \cap M(G)$ is annihilated by every $f \in \hat{T}$ with $|f| < 1$; that is, by every complex homomorphism of $\mathfrak{M}$ which annihilates $L(G)$. It follows that $L(G) \subset \mathfrak{N} \subset \text{Rad } L(G)$. This completes the proof of 7.1.4 and, hence, the proof of 7.1.3.

§7.6. A characterization of group algebras

7.6.1. One application of Theorem 7.1.3 is an abstract characterization of those measure algebras which are isomorphic (as measure algebras) to group algebras.

A class of measure algebras closely related to the group algebras is defined below.

DEFINITION. A measure algebra $\mathfrak{N}$ will be called an almost group algebra if there is an l. c. a. group G and an L-subalgebra $\mathfrak{N}' \subset M(G)$, with $L(G) \subset \mathfrak{N}' \subset \text{Rad } L(G)$, such that $\mathfrak{N}$ is isomorphic to $\mathfrak{N}'$ as a measure algebra.

7.6.2. PROPOSITION. *A measure algebra* $\mathfrak{N}$ *is an almost group algebra if and only if it is semisimple and its structure semigroup is a group.*

PROOF. If G is an l. c. a. group and $L(G) \subset \mathfrak{N} \subset \text{Rad}\,(L(G))$, then $L(G)$ and $\mathfrak{N}$ have the same spectrum and structure semigroup (3.5.3), namely $\overline{G}$ (4.2.1). Hence the structure semigroup of $\mathfrak{N}$ is a group. The converse follows directly from Theorem 7.1.3. In fact, 1 is trivially a critical point of $\hat{S}$ if S is a group.

7.6.3. GROUP ALGEBRAS. We shall call a measure algebra a group algebra if it is isomorphic, as a measure algebra, to $L(G)$ for some l. c. a. group G.

THEOREM. *A measure algebra is a group algebra if and only if it is semisimple and contains no nonzero proper L-ideals.*

PROOF. In 4.2.3 we proved that $L(G)$ contains no proper, nonzero L-ideals.

Conversely, suppose L is a semisimple measure algebra with no nonzero proper L-ideals. We claim that the structure semigroup S of L is a group. If it were not a group, we could find an element $f \in \hat{S}^+$ with $f < 1$. The ideal $J = \{s \in S : f(s) \leq \frac{1}{2}\}$ would then be the support of a nonzero proper L-ideal $L \cap M(J)$ of L. Now, by 7.6.2, L is an almost group algebra, and hence contains a group algebra as an L-ideal. Since L contains no nonzero proper L-ideals, it must be this group algebra.

7.6.4. COROLLARY. *A semisimple measure algebra $\mathfrak{N}$ is an almost group algebra if and only if it contains a minimal nonzero L-ideal L, L is a group algebra, and $\mathfrak{N}/L$ is a radical algebra.*

PROOF. Certainly if G is an l.c.a. group and $\mathfrak{N}$ an L-subalgebra of $M(G)$ with $L(G) \subset \mathfrak{N} \subset \mathrm{Rad}\,(L(G))$, then $\mathfrak{N}$ contains a minimal L-ideal, $L(G)$, and $\mathfrak{N}/L(G)$ is a radical algebra (every element has spectral radius zero).

Conversely, if $\mathfrak{N}$ contains a minimal L-ideal L, then L must be a group algebra by 7.6.3. If $\mathfrak{N}/L$ is a radical algebra, then $\mathfrak{N}$ and L have the same spectrum and hence the same structure semigroup. It follows from 7.6.2 that $\mathfrak{N}$ is an almost group algebra.

7.6.5. THE SPINE OF A MEASURE ALGEBRA. By the spine of a measure algebra $\mathfrak{M}$ we shall mean the closed linear span $\mathfrak{M}_0$ in $\mathfrak{M}$ of the group algebras in $\mathfrak{M}$.

Note that by 7.1.5, each group algebra in $\mathfrak{M}$ is contained in a maximal group algebra L and this maximal group algebra is the minimal ideal of the almost group algebra $\mathfrak{N} = \mathfrak{M} \cap M(G)$ for some maximal group $G \subset S$. Furthermore, since distinct maximal groups are disjoint, we have that $L_1 \perp L_2$ for distinct maximal group algebras L_1 and L_2. Thus $\mathfrak{M}_0$ is the direct sum of the maximal group algebras of $\mathfrak{M}$.

PROPOSITION. *The spine $\mathfrak{M}_0$ of a measure algebra $\mathfrak{M}$ is an L-subalgebra.*

PROOF. Let L_i be a maximal group algebra with support $G_i \subset S$ for $i = 1, 2$. If $p_1 \in G_1$ and $p_2 \in G_2$ are the idempotents, then $p_3 = p_1 p_2$ is the idempotent of some maximal group G_3 with $G_1 \cdot G_2 \subset G_3$. It follows that $L_1 * L_2$ is concentrated on G_3 and hence must be contained in the almost group algebra $\mathfrak{N}_3 = \mathfrak{M} \cap M(G_3)$. If L_3 is the group algebra which is the minimal ideal of $\mathfrak{N}_3$, then we claim that $L_1 * L_2 \subset L_3$.

Consider the map $\theta : L_1 \hat{\otimes} L_2 \to L_3$ induced by the multiplication map. This is an L-homomorphism. Furthermore, it follows from 2.5.2 that $L_1 \hat{\otimes} L_2$ is also a group algebra under the multiplication determined by $(\mu_1 \otimes \mu_2) * (\nu_1 \otimes \nu_2) = \mu_1 * \nu_1 \otimes \mu_2 * \nu_2$. The map θ is a homomorphism of measure algebras and its image is the closed linear span in $\mathfrak{N}_3$ of $L_1 * L_2$. Now it is easy to see that the image of a group algebra under such a homomorphism must also be a group algebra. Hence the image of θ contains no proper nonzero L-ideals. This implies that it is either contained in L_3 or meets it only in (0). The latter alternative is impossible, since $\mathfrak{N}_3/L_3$ is a radical algebra as is each of its subalgebras. We conclude that the product of any two maximal group algebras is contained in a third maximal group algebra and hence that $\mathfrak{M}_0$ is an algebra.

The following proposition follows directly from Theorem 7.1.6.

7.6.6. PROPOSITION. *If* $\mathfrak{M}$ *is a commutative measure algebra with spine* $\mathfrak{M}_0$ *and if* α *is the restriction map from the spectrum of* $\mathfrak{M}$ *to the spectrum of* $\mathfrak{M}_0$, *then* α *induces an isomorphism of cohomology for any cohomology sequence.*

7.6.7. AN INVOLUTION ON $\mathfrak{M}_0$. The subalgebra $\mathfrak{M}_0$ is concentrated on the sub-semigroup S_0 which is the union of the maximal groups in S. Recall from 3.4.7 that S_0 is closed in S. Furthermore, inversion in each maximal group defines a map $s \to s' : S_0 \to S_0$ which is a homeomorphism and satisfies $s'' = s$. We use this map to define an involution, $\mu \to \tilde{\mu}$, on $M(S_0)$ by setting

$$\tilde{\mu}(E) = \bar{\mu}(E').$$

PROPOSITION. *The spine,* $\mathfrak{M}_0$, *of* $\mathfrak{M}$ *is closed under the involution* $\mu \to \tilde{\mu}$ *and is symmetric* $((\tilde{\mu})\hat{} = (\hat{\mu})^-)$.

PROOF. When restricted to a maximal group algebra $L \subset \mathfrak{M}_0$, the involution $\mu \to \tilde{\mu}$ agrees with the natural involution in L. Also, L is symmetric under this involution. The proposition follows, since $\mathfrak{M}_0$ is the direct sum of its maximal group algebras.

CHAPTER 8. IDEMPOTENTS AND LOGARITHMS

In this chapter we apply the results on cohomology, obtained in the previous two chapters, to the study of the idempotents and invertible elements with logarithms in a measure algebra. We also apply the results on logarithms to the study of Wiener-Hopf operators on the line.

§8.1. The groups $H^0(\mathfrak{M})$ and $H^1(\mathfrak{M})$

8.1.1. Let $\mathfrak{M}$ be a commutative, semisimple measure algebra with a normalized identity δ and with structure semigroup S. As in Chapter 1, we let $H^0(\mathfrak{M})$ be the additive subgroup of $\mathfrak{M}$ generated by the idempotents of $\mathfrak{M}$ and $H^1(\mathfrak{M})$ the group $\mathfrak{M}^{-1}/\exp(\mathfrak{M})$.

Recall that the Gelfand transform $\mu \to \hat{\mu}$ induces isomorphisms $H^0(\mathfrak{M}) \to H^0(\hat{S})$ and $H^1(\mathfrak{M}) \to H^1(\hat{S})$ (cf. 1.4.2 and 1.6.2).

8.1.2. Algebras Without Identity. In case $\mathfrak{M}$ does not have an identity, $\hat{S}$ is only locally compact. Recall that the groups $H^0_c(\hat{S})$ and $H^1_c(\hat{S})$ are defined by passing to the one-point compactification $\hat{S} \cup \{\infty\}$ and then identifying $H^p_c(\hat{S})$ with the canonical image of $H^p(\hat{S} \cup \{\infty\}, \{\infty\})$ in $H^p(\hat{S} \cup \{\infty\})$.

It is apparent that $H^0_c(\hat{S})$ consists of those continuous integer valued functions on $\hat{S} \cup \{\infty\}$ which vanish at $\{\infty\}$, i.e., the integer valued elements of $C_0(\hat{S})$. It follows that the Gelfand transform $\mu \to \hat{\mu} : \mathfrak{M} \to C_0(\hat{S})$ is an isomorphism of $H^0_c(\mathfrak{M})$ onto $H^0_c(\hat{S})$ if we let $H^0_c(\mathfrak{M})$ be the group generated by the idempotents in $\mathfrak{M}$.

The group $H^1_c(\hat{S})$ is just the subgroup of $H^1(\hat{S} \cup \{\infty\})$ determined by elements of $C(\hat{S} \cup \{\infty\})^{-1}$ which have the value one at infinity. However, if $f \in C(\hat{S} \cup \{\infty\})^{-1}$ and $z = \ln f(\infty)$, $g = fe^{-z}$, then f and g determine the same equivalence class in $H^1(\hat{S} \cup \{\infty\})$. Thus, $H^1_c(\hat{S}) = H^1(\hat{S} \cup \{\infty\})$.

Thus, if we set $H^1_c(\mathfrak{M}) = H^1(\mathfrak{M} + \mathbf{C}\delta)$, where $\mathfrak{M} + \mathbf{C}\delta$ is the algebra obtained by adjoining a normalized identity, then $H^1_c(\mathfrak{M})$ and $H^1_c(\hat{S})$ are isomorphic via the Gelfand transform.

If $\mathfrak{M}$ has an identity and $\mathfrak{N} \subset \mathfrak{M}$ is a subalgebra (possibly not containing the identity), then the embedding of $\mathfrak{N} + \mathbf{C}\delta$ in $\mathfrak{M}$ obviously induces natural maps $H^0_c(\mathfrak{N}) \to H^0(\mathfrak{M})$ and $H^1_c(\mathfrak{N}) \to H^1(\mathfrak{M})$.

8.1.3. Theorem. *Let $\mathfrak{M}$ be a semisimple commutative measure algebra with a normalized identity. Then for each maximal group $L_h \subset \mathfrak{M}$, the maps $H^p_c(L_h) \to H^p(\mathfrak{M})$*

$(p = 0, 1)$ *are injective. Furthermore,* $H^p(\mathfrak{M})$ $(p = 0, 1)$ *is the direct sum of the images of these maps as* L_h *ranges over all maximal group algebras in* $\mathfrak{M}$.

PROOF. Let S be the structure semigroup of $\mathfrak{M}$. For each critical point $h \in \hat{S}^+$, let L_h be the corresponding maximal group algebra and Γ_h its spectrum.

Now the restriction map $f \to f|_{L_h}$ maps the spectrum of $\mathfrak{M}$ onto $\Gamma_h \cup \{\infty\}$. Since $\{H^0, H^1\}$ is a cohomology sequence (6.2.7), it follows from Theorem 7.1.6 that the corresponding map $H_c^p(\Gamma_h) \to H^p(\hat{S})$ of cohomology is injective and $H^p(\hat{S})$ is the direct sum of the images as h ranges over all critical points. The theorem now follows from the fact that the diagram

$$\begin{array}{ccc} H_c^p(L_h) & \longrightarrow & H^p(\mathfrak{M}) \\ \downarrow & & \downarrow \\ H_c^p(\Gamma_h) & \longrightarrow & H^p(\hat{S}) \end{array}$$

is commutative and the vertical maps are isomorphisms.

8.1.4. IDEMPOTENTS IN GROUP ALGEBRAS. Let G be an l.c.a. group and $\nu \in L(G)$ an idempotent. Then $\hat{\nu}$ is the characteristic function of an open-compact set $U \subset \hat{G}$. If $\gamma \in U$ then $\gamma^{-1}U$ must contain the connected component Δ of the identity in $\hat{G}$. Since $\hat{G}/\Delta$ is totally disconnected, it contains open-compact subgroups in every neighborhood of the identity. It follows that $\gamma^{-1}U$ contains some open-compact subgroup of $\hat{G}$.

By using the above argument for each $\gamma \in U$ and the fact that U is compact, we may conclude that U is the union of finitely many cosets of open-compact subgroups of $\hat{G}$.

Now if V is an open-compact subgroup of $\hat{G}$, then $\chi_V = \hat{\mu}$, where μ is the normalized Haar measure of the compact group $V^\perp = \{g \in G : \gamma(g) = 1 \text{ for } \gamma \in V\}$. For a coset γV of V the characteristic function is $(\gamma\mu)^\wedge$.

It follows from the above considerations that:

PROPOSITION. *Each idempotent of* $L(G)$ *is in the group generated by idempotents of the form* $\gamma\mu$ *for* $\gamma \in G$ *and* μ *the normalized Haar measure of some open-compact subgroup of* G.

In view of this, we have the following corollary of Theorem 8.1.3.

8.1.5. COROLLARY. *Let* $\mathfrak{M}$ *be a commutative, semisimple measure algebra. Then each idempotent in* $\mathfrak{M}$ *is in the group generated by measures of the form* $\gamma\mu$, *where for some critical point* $h \in \hat{S}^+$, $\gamma \in \Gamma_h$, *and* $\mu \in L_h$ *is the image of the normalized Haar measure on an open-compact subgroup of* G_h *under the isomorphism* $L(G_h) \to L_h$.

When phrased in concrete terms, Theorem 8.1.3 applied to $H^1(\mathfrak{M})$ yields the following:

8.1.6. COROLLARY. *Let* $\mathfrak{M}$ *be a commutative, semisimple measure algebra with normalized identity. If* $\mu \in \mathfrak{M}^{-1}$ *then there are distinct maximal group algebras* $L_{h_1}, \cdots, L_{h_n}$, *elements* $\nu_i \in (L_{h_i} + C\delta)^{-1}$, *for* $i = 1, \cdots, n$, *and an element* $\omega \in \mathfrak{M}$, *such that*

$$\mu = \nu_1 * \nu_2 * \cdots * \nu_n * e^{\omega}.$$

Furthermore, each ν_i *is unique modulo* $\exp(L_{h_i} + \mathbf{C}\delta)$.

§8.2. The algebra $M(G)$

8.2.1. We now interpret the above results in the case of the measure algebra $M(G)$ on an l. c. a. group G. The first task is to identify the maximal group algebras of $M(G)$.

It follows from the discussion in 4.3.5 that $L(G_\tau)$ is a maximal group algebra in $M(G)$ for each l. c. a. group G_τ which is continuously isomorphic to G. Conversely, we have

8.2.2. PROPOSITION. *If* L *is a maximal group algebra in* $M(G)$, *then* $L = L(G_\tau)$ *for some l. c. a. group* G_τ *continuously isomorphic to* G.

PROOF. Let G' be an l. c. a. group and $\phi : L(G') \to L \subset M(G)$ an isomorphism of measure algebras. Then, as in 3.3.3, the adjoint ϕ^* of ϕ maps the spectrum $\hat{S}$ of $M(G)$ continuously into $\hat{G}' \cup \{\infty\}$. Furthermore, this map preserves pointwise products.

Now, no element of $\hat{G}$ restricts to the zero homomorphism of L. Hence ϕ^* is a continuous group homomorphism of $\hat{G}$ into $\hat{G}'$. It follows that it is induced by a continuous group homomorphism $\alpha : G' \to G$. Also, the identity

$$\int \phi^*\gamma \, d\mu = \int \gamma \circ \alpha \, d\mu = \int \gamma \, d(\mu \circ \alpha^{-1}),$$

for $\gamma \in G$, $\mu \in L(G')$, implies that $\phi\mu = \mu \circ \alpha^{-1}$.

The map $\mu \to \mu \circ \alpha^{-1} : L(G') \to M(G)$ cannot be an isometry unless α is an isomorphism onto a subgroup G'' of G. Thus, we may construct a new l. c. a. group topology τ on G by declaring G'' to be open and giving G'' the topology for which α is a homeomorphism. The resulting l. c. a. group G_τ has the property that L is just the subalgebra of $L(G_\tau)$ consisting of measures concentrated on the open subgroup G''_τ. Since L is maximal, we have $G'' = G$ and $L = L(G_\tau)$.

Thus, in the case where $\mathfrak{M} = M(G)$, 8.1.5 and 8.1.6 yield the following results.

8.2.3. THEOREM (COHEN [1]). *Each idempotent in* $M(G)$ *is in the additive group generated by the idempotents of the form* $\gamma\mu$, *where* $\gamma \in \hat{G}$ *and* μ *is the Haar measure on some compact subgroup of* G.

PROOF. We need only notice that if G_τ is continuously isomorphic to G, then a compact subgroup of G_τ is also a compact subgroup of G, and that a character on a compact subgroup is the restriction of a character on G.

8.2.4. THEOREM. *For each* $\mu \in M(G)^{-1}$ *there are l. c. a. groups* $G_{\tau_1}, \cdots, G_{\tau_n}$ *continuously isomorphic to* G *and measures* $\nu_i \in (L(G_{\tau_i}) + \mathbf{C}\delta)^{-1}$, $\omega \in M(G)$, *such that*

$$\mu = \nu_1 * \cdots * \nu_n * e^{\omega}.$$

The measures ν_i *are unique modulo* $\exp(L(G_{\tau_i}) + \mathbf{C}\delta)$.

§8.3. Logarithms in group algebras

8.3.1. By 7.1.6 we have reduced the problem of computing the cohomology of the spectrum of a measure algebra to the corresponding problem for each of its maximal group algebras. That is, to compute $H^p(\hat{S})$ we need only identify the maximal group algebras $L_h \subset \mathfrak{M}$ and then compute $H^p_c(\Gamma_h)$ for the spectrum Γ_h of each such group algebra.

Now each l. c. a. group Γ has an open subgroup of the form $\Gamma_0 \times R^n$ with Γ_0 compact. Hence, Γ is the discrete union of cosets of $\Gamma_0 \times R^n$. It follows that we can compute $H^p_c(\Gamma)$ if we can compute $H^p_c(\Gamma_0 \times R^n)$.

The one-point compactification of $\Gamma_0 \times R^n$ is $\Gamma_0 \times S^n$, where $S^n = R^n \cup \{\infty\}$ is an n-sphere. If we are dealing with Čech cohomology with coefficients in a group K, then $H^p_c(R^n) = H^p(R^n \cup \{\infty\}, \{\infty\}) = K$ if $p = n$ and $H^p_c(R^n) = 0$ otherwise.

The Künneth formula (Bredon [1]) now allows us to compute $H^p_c(\Gamma_0 \times R^n)$ if we know $H^p_c(\Gamma_0) = H^p(\Gamma_0)$. Thus the problem of computing the cohomology (with compact supports) of an l. c. a. group can be reduced to the corresponding problem for compact groups. This latter problem is solved in Hofmann [1]. Here we shall only discuss the special cases $p = 0, 1$ and $K = Z$ which are important in the study of idempotents and logarithms.

The groups $H^0_c(\Gamma)$ were computed in 8.1.4.

8.3.2. THE GROUP $H^1(\Gamma)$ FOR COMPACT, CONNECTED Γ. Let Γ be a compact connected group. We shall compute the group $C(\Gamma)^{-1}/\exp(C(\Gamma)) = H^1(\Gamma)$.

Now it follows from the homotopy axiom (which is satisfied by H^1, according to 6.2.6) that the translation maps $\gamma_1 \to \gamma\gamma_1 : \Gamma \to \Gamma$ induce automorphisms T_γ of $H^1(\Gamma)$ which are locally constant functions of γ. Since Γ is connected and $T_1 = \text{id}$, the translations must all act trivially on $H^1(\Gamma)$. Thus, if $f \in C(\Gamma)^{-1}$ then $f(\gamma\gamma_1) = f(\gamma_1)e^{\phi(\gamma,\gamma_1)}$ for some function ϕ which is continuous in γ_1 for each γ.

Note that $f(\gamma) = f(1)e^{\phi(\gamma,1)}$. We can normalize f so that $f(1) = 1$ without changing its class in $H^1(\Gamma)$. If we do this, then

(i) $$f(\gamma\gamma_1) = f(\gamma)f(\gamma_1)e^{h(\gamma,\gamma_1)},$$

where $h(\gamma, \gamma_1) = \phi(\gamma, \gamma_1) - \phi(\gamma, 1)$. Note that h is continuous as a function of γ_1 for each γ and may be chosen to be a Borel function on $\Gamma \times \Gamma$.

Now given $\gamma, \gamma_1, \gamma_2 \in \Gamma$ we have

$$f(\gamma\gamma_1\gamma_2) = f(\gamma)f(\gamma_1\gamma_2)e^{h(\gamma,\gamma_1\gamma_2)} = f(\gamma)f(\gamma_1)f(\gamma_2)e^{h(\gamma_1,\gamma_2)+h(\gamma,\gamma_1\gamma_2)}$$

but also

$$f(\gamma\gamma_1\gamma_2) = f(\gamma\gamma_1)f(\gamma_2)e^{h(\gamma\gamma_1,\gamma_2)} = f(\gamma)f(\gamma_1)f(\gamma_2)e^{h(\gamma,\gamma_1)+h(\gamma\gamma_1,\gamma_2)}.$$

Hence

(ii) $$h(\gamma_1, \gamma_2) = h(\gamma, \gamma_1) + h(\gamma\gamma_1, \gamma_2) - h(\gamma, \gamma_1\gamma_2) \mod (2\pi i).$$

We set $k(\gamma_1) = \int h(\gamma, \gamma_1)\, dm(\gamma)$, where m is normalized Haar measure on Γ. On integrating equation (ii) over Γ we obtain

(iii) $$h(\gamma_1, \gamma_2) = k(\gamma_1) + k(\gamma_2) - k(\gamma_1 \gamma_2) \mod (2\pi i).$$

Note that k is continuous. Hence f is equivalent mod exp $C(\Gamma)$ to $l = fe^k$. However, by (i) and (ii) we have

$$l(\gamma_1 \gamma_2) = l(\gamma_1)\, l(\gamma_2).$$

Hence l is a continuous character on Γ and must have the form $(\delta_g)^\wedge$ for some $g \in G$, where $\hat{G} = \Gamma$.

Now no character on Γ, other than 1, can have a continuous logarithm. In fact, a continuous logarithm for an element of $\hat{\Gamma}$ would be a continuous homomorphism of the compact group Γ into R. Its image would have to be a compact subgroup of R and hence would have to be (0).

We conclude:

PROPOSITION. *If* Γ *is a compact connected group, then* $H^1(\Gamma) = C(\Gamma)^{-1}/\exp(\Gamma)$ *is isomorphic to the dual group of* Γ. *That is, each* $f \in C(\Gamma)^{-1}$ *has a factorization as* $f = le^h$ *for a unique* $l \in \hat{\Gamma}$ *and some* $h \in C(\Gamma)$.

As a corollary we have the following result which is due to Bohr [1] in the case of the line.

8.3.3. COROLLARY. *If* G *is a discrete torsion free group then each* $\mu \in L(G)^{-1}$ *has a factorization as*

$$\mu = \delta_g e^\nu$$

for some $\nu \in L(G)$ *and a unique element* $g \in G$.

PROOF. Since G is discrete and torsion free, its dual group $\hat{G} = \Gamma$ is compact and connected (Rudin [5]).

The characters on Γ are exactly the Fourier transforms of the point masses δ_g for $g \in G$. Hence, the corollary follows directly from 8.3.2 and the Arens-Royden Theorem 1.6.2.

§8.4. Logarithms in $M(R)$

8.4.1. We are now in a position to compute $M(R)/\exp(M(R))$.

Note that the groups continuously isomorphic to R are just R and R_d. Hence, by 8.2.4, each $\mu \in M(R)^{-1}$ has the form $\mu = \nu_1 * \nu_2 * e^\omega$ for $\nu_1 \in (L(R) + \mathbf{C}\delta)^{-1}$ and $\nu_2 \in L(R_d)^{-1}$.

Now, by 1.6.4, $\nu_1 = N^k * e^{\omega_1}$ for a unique integer k, where $\hat{N}(y) = (1 + iy)(1 - iy)^{-1}$. Similarly, $\nu_2 = \delta_c * e^{\omega_2}$ for a unique $c \in R$, by 8.3.3. Hence,

THEOREM. *If* $\mu \in M(R)^{-1}$ *then* μ *has a factorization as*

$$\mu = N^k * \delta_c * e^\omega,$$

where k *is a unique integer,* c *a unique real number, and* $N \in (L(R) + \mathbf{C}\delta)^{-1}$ *is the measure with Fourier transform* $\hat{N}(y) = (1 + iy)(1 - iy)^{-1}$.

Note that this theorem says that $H^1(M(R)) = M(R)^{-1}/\exp(M(R))$ is isomorphic to $Z \times R$, as is $H^1(\hat{S})$ if S is the structure semigroup of $M(R)$.

8.4.2. COROLLARY. *Let* $B(R) = \{f \in C(R) : f = \hat{\mu}$ *for some* $\mu \in M(R)\}$. *Then if* f *and* f^{-1} *are elements of* $B(R)$, *we have that* $\ln f(y) = k \ln [(1+iy)(1-iy)^{-1}] - icy + g(y)$ *for a unique integer* k, *a unique number* c, *and some function* $g \in B(R)$. *Here,* $\ln f$ *is any continuous logarithm for* f *on* R.

8.4.3. COROLLARY. *If* $f \geqslant 0, f, f^{-1} \in B(R)$, *then* $\ln f \in B(R)$.

8.4.4. COMPUTING THE INDICES k AND c. For an invertible measure $\mu \in M(R)$, we call the numbers $k = k(\mu)$ and $c = c(\mu)$ of 8.4.1 the discrete and continuous indices of μ, respectively. They can be computed as follows: by 8.4.2, $c(\mu)$ is the unique real number c such that $\ln \hat{\mu}(y) + icy$ is bounded on R. Furthermore,

$$\text{(i)} \qquad \ln \hat{\mu}(y) + ic(\mu)y = k(\mu) \ln [(1+iy)(1-iy)^{-1}] + g(y)$$

for some $g \in B(R)$.

Now the numbers λ_+ and λ_- given by

$$\lambda_\pm(f) = \lim_{N \to \pm\infty} \frac{1}{N} \int_0^N f(y)\,dy$$

must be equal if $f \in B(R)$. In fact, if $f = \hat{\nu}$ for $\nu \in M(R)$, then $\lambda_+ = \lambda_- = \nu(\{0\})$. However, for $f(y) = \ln [(1+iy)(1-iy)^{-1}]$, the numbers $\lambda_+(f)$ and $\lambda_-(f)$ are given by

$$\lambda_\pm(f) = \lim_{y \to \pm\infty} f(y)$$

and, hence, $\lambda_+ - \lambda_- = 2\pi i$. It follows that the number $k(\mu)$ is $(1/2\pi i)\,[\lambda_+(f) - \lambda_-(f)]$ for the function $f(y) = \lim \hat{\mu}(y) + ic(\mu)y$ of equation (i).

§8.5. Weiner-Hopf equations

8.5.1. Let $\mu \in M(R)^{-1}$ have the factorization $\mu = N^k * \delta_c * e^\omega$ as in 8.4.1. We set $\omega_+ = \omega|_{R^+}$ and $\omega_- = \omega - \omega_+$.

Now the measure N is concentrated on R^-. In fact, $N = 2gm - \delta$, where $g(x) = \chi_{[-\infty,0]}(x)e^x$. Similarly, N^{-1} is concentrated on R^+. We also have $\delta_c \in M(R^+)$ if $c \geqslant 0$ and $\delta_c \in M(R^-)$ if $c \leqslant 0$. Thus we can always arrange the factors in $\mu = N^k * \delta_c * e^{\omega_+} * e^{\omega_-}$ so that those in $M(R^+)$ are on the right and those in $M(R^-)$ are on the left. Thus it follows from Lemma 1.5.3 that W_μ is a product, in some order, of the Weiner-Hopf operators $W_{N^k}, W_{\delta_c}, W_{e^{\omega_+}}, W_{e^{\omega_-}}$.

Now W_{δ_c} is never invertible unless $c = 0$. In fact, for $c > 0$, W_{δ_c} has a proper image $L^p[c, \infty)$ in $L^p(R^+)$, while, for $c < 0$, W_{δ_c} has a nonzero kernel, $L^p[0, \bar{c}]$. Thus, if $c \neq 0$ then W_{δ_c} has either an infinite dimensional kernel or an infinite dimensional cokernel.

The operator W_N is surjective but has a one dimensional kernel, while $W_{N^{-1}}$ is injective with a one dimensional cokernel. To see this, note that $W_{N^{-1}}f = N^{-1} * f$ for $f \in L^p(R^+)$. Since $(N^{-1})^\wedge$ is analytic on $C^+ = \{z \in \mathbf{C} : \mathrm{Im}(z) > 0\}$, we conclude that

$f \to N^{-1} * f$ is injective. Now for $g \in L^p(R^+)$, $g \in$ image (W_{N-1}) if and only if g vanishes at i, we conclude that W_{N-1} has a one dimensional cokernel on $L^p(R^+)$. Since W_N is a left inverse for W_{N-1}, we conclude that it is surjective and has a one dimensional kernel.

Now the operators $W_{e^{\omega_+}}$ and $W_{e^{\omega_-}}$ are invertible by 1.5.4. Hence, it follows from the above that a product of W_{δ_c}, W_{N^k}, $W_{e^{\omega_+}}$, and $W_{e^{\omega_-}}$ will have an infinite dimensional kernel or cokernel unless $c = 0$. If $c = 0$ then such a product will have a finite dimensional (but nonzero) kernel or cokernel unless $k = 0$. Hence

PROPOSITION. *If $\mu \in M(R)^{-1}$ then, for any p, the Weiner-Hopf operator W_μ is invertible on $L^p(R^+)$ if and only if $\mu = e^\omega$ for some $\omega \in M(R)$. This happens if and only if the indices $k(\mu)$ and $c(\mu)$ are both zero.*

8.5.2. THE CASE $p = 1$. It can happen that, for a measure $\mu \in M(R)$, the operator W_μ is invertible on $L^2(R^+)$ even though μ is not invertible in $M(R)$ (Douglas and Taylor [1]). We shall show that this cannot happen if $p = 1$.

Suppose $\mu \in M(R)$ and W_μ is invertible. Then if $M = \| W_\mu^{-1} \|$, we have

$$\| \mu * f \| \geqslant \| (\mu * f)|_{R^+} \| = \| W_\mu f \| \geqslant (1/M) \| f \|$$

for each $f \in L^1(R^+)$. However, if $f \in L^1[-x, \infty)$ for $x > 0$, then $\delta_x * f \in L^1(R^+)$ and

$$\| \mu * f \| = \| \mu * \delta_x * f \| \geqslant (1/M) \| \delta_x * f \| = (1/M) \| f \|.$$

It follows that $\| f \| \leqslant M \| \mu * f \|$ for every function in $L^1(R)$ with compact support and, hence, for every function in $L^1(R)$. We conclude that the convolution operator $C_\mu : L^1(R) \to L^1(R)$ is injective and has closed range. It follows from Rudin [5, 2.6.3] that C_μ must be surjective and, hence, invertible. By 1.2.8, μ is invertible in $M(R)$. Thus,

THEOREM (DOUGLAS AND TAYLOR [1]). *If $\mu \in M(R)$, then the Weiner-Hopf operator W_μ is invertible on $L^1(R^+)$ if and only if μ has the form $\mu = e^\omega$ for some $\omega \in M(R)$, that is, if and only if $\mu \in M(R)^{-1}$ and $k(\mu) = c(\mu) = 0$.*

8.5.3. REMARK. In order to apply the preceding results for a given $\mu \in M(R)$, one needs to do two things: (i) decide whether or not μ is invertible in $M(R)$, and (ii) compute the numbers k and c for the factorization of μ. The first problem is the subject of the next chapter. The second is easy in view of 8.4.4.

§8.6. One parameter groups in $M(R)$

8.6.1. If C is an invertible operator acting on a Banach space X, then a question of some interest is whether or not C can be embedded in a strongly continuous one parameter group of operators on X. That is, is there a group homomorphism $t \to \alpha(t)$ from R into the group of invertible operators on X such that $t \to \alpha(t)x$ is continuous for each x and $\alpha(1) = C$?

Fisher [1] uses the factorization 8.4.1 to prove that if $\mu \in M(R)^{-1}$, then the convolution operator $C_\mu : L^1(R) \to L^1(R)$ is $\alpha(1)$ for a strongly continuous group $t \to \alpha(t)$

of convolution operators. Theorem 8.4.1 allows Fisher to reduce the general problem to the case of the measure N of 8.4.1. In fact if $\mu = N^k * \delta_c * e^\omega$ then, since e^ω and δ_c are obviously embeddable in strongly continuous one-parameter groups, μ will be so embeddable if N is.

We shall not reconstruct Fisher's proof of the embeddability of N here. It involves proving that $t \to (1 - D)^t (1 + D)^{-t}$ is a strongly continuous group of convolution operators $(D = d/dt)$ and noting that $C_N = (1 - D)(1 + D)^{-1}$.

CHAPTER 9. INVERTIBLE MEASURES

The results of the previous two chapters can be used to significantly simplify the problem of deciding when an element of a measure algebra is invertible. The result we shall prove here implies that in seeking an inverse for $\mu \in \mathfrak{M}$, one need only search in a certain "small" subalgebra of $\mathfrak{M}$ containing μ.

§9.1. The main theorem

9.1.1. BALANCED L-SUBALGEBRAS. Let $\mathfrak{M}$ be a semisimple commutative measure algebra with a norm one approximate identity. As usual, S will denote the structure semigroup of $\mathfrak{M}$. For each critical point h, L_h will denote the corresponding maximal group algebra.

Now L-subalgebras of group algebras come in two types: those which are group algebras themselves and those which are not. If $\mathfrak{N}$ is an L-subalgebra of $L(G)$, for an l.c.a. group G, then the smallest closed set T which supports $\mathfrak{N}$ is a subsemigroup of G and $\mathfrak{N} = L(T) = L(m\,|_T)$. Clearly, $\mathfrak{N}$ is a group algebra if and only if T is a subgroup (4.5.2). In this case, T must be an open subgroup (otherwise, $m|_T = 0$). The subsemigroup T is a group if and only if it is closed under inversion, and this happens if and only if $\mathfrak{N}$ is closed under the involution $\mu \to \tilde{\mu}$.

Returning to our measure algebra $\mathfrak{M}$, we shall be interested in L-subalgebras $\mathfrak{N}$ which satisfy the following condition.

DEFINITION. The L-subalgebra $\mathfrak{N} \subset \mathfrak{M}$ will be called balanced if for every maximal group algebra L_h either $\mathfrak{N} \cap L_h$ is zero or it is a subgroup algebra of L_h.

9.1.2. THEOREM. *Let $\mathfrak{M}$ be a semisimple commutative measure algebra with normalized identity and let $\mathfrak{N}$ be a balanced L-subalgebra of $\mathfrak{M}$ containing the identity. Then for each $\mu \in \mathfrak{N}$, μ has an inverse in $\mathfrak{N}$ if and only if it has an inverse in $\mathfrak{M}$.*

We shall prove this through a sequence of lemmas. Recall that the spine of $\mathfrak{M}$ is the L-subalgebra $\mathfrak{M}_0 = \oplus_h L_h$ (7.6.5).

9.1.3. LEMMA. *If $\mathfrak{N}$ is a balanced L-subalgebra of $\mathfrak{M}$, then the spine $\mathfrak{N}_0$ of $\mathfrak{N}$ is $\mathfrak{M}_0 \cap \mathfrak{N}$. Furthermore, each complex homomorphism of $\mathfrak{N}_0$ extends to a complex homomorphism of $\mathfrak{M}$.*

PROOF. That $\mathfrak{N}_0 = \mathfrak{M}_0 \cap \mathfrak{N}$ is exactly the statement that $\mathfrak{N}$ is balanced.

Now $\mathfrak{N}_0$ has an involution under which it is a symmetric algebra (7.6.7). It follows that the Shilov boundary of $\mathfrak{N}_0$ is its spectrum and complex homomorphisms extend.

9.1.4. LEMMA. *Each* $\mu \in \mathfrak{M}^{-1}$ *has a factorization as* $\mu = \nu * e^{\omega}$ *for a unique pair* ν, ω *with* $\nu \in \mathfrak{M}_0^{-1}$ *and* $\omega \in \mathfrak{M}_0^{\perp}$.

PROOF. According to 8.1.6 we may write μ as $\nu_1 * \cdots * \nu_n * e^{\omega_1}$ with $\nu_1, \cdots, \nu_n \in \mathfrak{M}_0^{-1}$ and $\omega_1 \in \mathfrak{M}$. If $\omega_1 = \omega + \omega_2$ with $\omega \in \mathfrak{M}_0^{\perp}$ and $\omega_2 \in \mathfrak{M}_0$, then $\mu = \nu * e^{\omega}$ for $\nu = \nu_1 * \cdots * \nu_n * e^{\omega_2} \in \mathfrak{M}_0^{-1}$.

Suppose now that $\mu = \nu' * e^{\omega'}$ is another factorization of μ with $\nu' \in \mathfrak{M}_0^{-1}$ and $\omega' \in \mathfrak{M}_0^{\perp}$. Then $e^{\omega - \omega'} \in \mathfrak{M}_0$. However, by 7.6.6, the groups $H^1(\mathfrak{M})$ and $H^1(\mathfrak{M}_0)$ are isomorphic. Hence, we must have $e^{\omega - \omega'} = e^{\omega''}$ for some $\omega'' \in \mathfrak{M}_0$. Then $(2\pi i)^{-1}(\omega - \omega' - \omega'')$ is an element of the group $H^0(\mathfrak{M}) = H^0(\mathfrak{M}_0)$. Thus, $\omega - \omega' - \omega'' \in \mathfrak{M}_0$. Since $\omega - \omega' \in \mathfrak{M}_0^{\perp}$ and $\omega'' \in \mathfrak{M}_0$, we conclude that $\omega - \omega' = 0$ and hence $\omega = \omega'$ and $\nu = \nu'$.

9.1.5. LEMMA. *Let* $\mathfrak{N}$ *be a balanced L-subalgebra of* $\mathfrak{M}$ *containing the identity. If* $\mu \in \mathfrak{N}^{-1}$ *and* $\mu = \nu * e^{\omega}$ *is the factorization of* 9.1.4, *then* $\nu \in \mathfrak{N}^{-1}$ *and* $\omega \in \mathfrak{N}$.

PROOF. The hypothesis that $\mathfrak{N}$ is balanced means exactly that the spine of $\mathfrak{N}$ is the intersection of $\mathfrak{N}$ with the spine of $\mathfrak{M}$, that is, $\mathfrak{N}_0 = \mathfrak{M}_0 \cap \mathfrak{N}$. Hence, if $\mu = \nu_1 * e^{\omega_1}$ is the factorization of 9.1.4 with $\nu_1 \in \mathfrak{N}_0^{-1}$ and $\omega_1 \in \mathfrak{N}_0^{\perp}$, then $\nu_1 \in \mathfrak{M}_0$ and $\omega_1 \in \mathfrak{M}_0^{\perp}$ as well. Hence, by the uniqueness of the factorization, $\nu_1 = \nu$ and $\omega_1 = \omega$.

9.1.6. LEMMA. *Let* $\mathfrak{N}$ *be a balanced L-subalgebra of* $\mathfrak{M}$ *containing the identity. Let* μ *be an element of* $\mathfrak{N} \cap \mathfrak{M}^{-1}$. *If* $f\mu \in \mathfrak{N}^{-1}$ *for every* $f \in \hat{S}^+$ *with* $f < 1$, *then* $\mu \in \mathfrak{N}^{-1}$.

PROOF. This is obvious unless 1 is a critical point. Thus, assume 1 is a critical point and let L and Rad L be the corresponding group algebra and almost group algebra. Note that if K is the kernel of S, then Rad $L = \mathfrak{M} \cap M(K) = \{\nu \in \mathfrak{M} : f\nu = 0$ for all $f \in \hat{S}^+$ with $f < 1\}$.

Suppose that $L \cap \mathfrak{N} = (0)$. Then $|\mu|^n \perp L$ for every n and hence the image of $|\mu|$ in $\mathfrak{M}/L$ has spectral radius $\|\mu\|$. The spectrum of $\mathfrak{M}/L$ is $\{f \in \hat{S} : \hat{\nu}(f) = 0$ for all $\nu \in L\} = \{f \in \hat{S} : |f| < 1\}$. It follows that there exists f in this set with $||\mu|\,\hat{}\,(f)| = \|\mu\|$. This implies that $|f| = 1$ a.e./μ and hence $|f|\mu = \mu$. Since $|f| < 1$, we conclude from the hypothesis that $\mu \in \mathfrak{N}^{-1}$.

Now suppose that $L \subset \mathfrak{N}$. We use the factorization $\mu = \nu * e^{\omega}$ $(\nu \in \mathfrak{M}_0, \omega \in \mathfrak{M}_0^{\perp})$ of 9.1.4. Note that by 9.1.5 we have $f\nu \in \mathfrak{N}^{-1}$ and $f\omega \in \mathfrak{N}$ for each $f \in \hat{S}^+$ with $f < 1$. It follows that if $\nu = \nu_1 + \nu_2$, $\omega = \omega_1 + \omega_2$ with $\nu_1, \omega_1 \in$ Rad L, $\nu_2, \omega_2 \in$ (Rad $L)^{\perp}$, then $\nu_2, \omega_2 \in \mathfrak{N}$. Furthermore, $\nu_1 \in \mathfrak{M}_0 \cap$ (Rad L) $= L \subset \mathfrak{N}$. Hence, $\nu \in \mathfrak{N} \cap \mathfrak{M}_0 = \mathfrak{N}_0$ and, by 9.1.3, $\nu^{-1} \in \mathfrak{N}_0$. It follows that $e^{\omega_2} = \mu * \nu^{-1} * e^{-\omega_1} \in \mathfrak{N} \cap$ Rad L. Since $L \subset \mathfrak{N} \cap$ Rad $L \subset$ Rad L, we conclude that $\mathfrak{N} \cap$ Rad L and Rad L have the same spectrum and hence that e^{ω_2} (which has an inverse $e^{-\omega_2}$ in Rad L) must have an inverse in $\mathfrak{N} \cap$ Rad L. Hence $\mu^{-1} = \nu^{-1} * e^{-\omega_1} * e^{-\omega_2} \in \mathfrak{N}$.

We complete the proof by showing that if $L \cap \mathfrak{N} \neq (0)$, then we can reduce to the case where $L \subset \mathfrak{N}$. Note that if $L \cap \mathfrak{N} \neq (0)$, then it is a group algebra and an L-ideal of $\mathfrak{N}$. If $L \simeq L(G)$ then, as in §7.4, we may pass to a representation of $\mathfrak{M}$ on a locally compact semigroup T which has G as kernel. Then $L = L(G)$ and $L \cap \mathfrak{N} = L(G_1)$ for some open subgroup $G_1 \subset G$. Let $p \in G$ be the idempotent and set $\mathfrak{M}_1 = \mathfrak{M} \cap M(T_1)$ where $T_1 = \{t \in T : pt \in G_1\}$. Note that since $L \cap \mathfrak{N}$ is an ideal of $\mathfrak{N}$, we have $\mathfrak{N} \subset \mathfrak{M}_1$. Also, by 4.5.6, every complex homomorphism of $\mathfrak{M}_1$ extends to $\mathfrak{M}$. Hence, $\mu^{-1} \in \mathfrak{M}_1$. Now in $\mathfrak{M}_1$ the group algebra corresponding to 1 (i.e., the minimal L-ideal of $\mathfrak{M}_1$) is just $L(G_1) = L \cap \mathfrak{N}$. Hence, we have reduced the problem to the previous case, and the proof is complete.

9.1.7. Proof of Theorem 9.1.2. Let $\mu \in \mathfrak{N}$ have an inverse μ^{-1} in $\mathfrak{M}$. Then the strongly closed set $\{f \in \hat{S}^+ : f\mu^{-1} \in \mathfrak{N}\}$ and the strongly open set $\{f \in \hat{S}^+ : f\mu \in \mathfrak{N}^{-1}\}$ are identical ($f\mu^{-1} = (f\mu)^{-1}$). It follows from 5.1.6 and 5.2.4 that either this set is all of $\hat{S}^+$ (and hence it contains 1) or else its complement contains a minimal element h which is a critical point. The latter alternative is impossible since $h\mu$ would then satisfy the hypothesis of 9.1.6 with $\mathfrak{N}$ and $\mathfrak{M}$ replaced by $\mathfrak{N}_h = \mathfrak{N} \cap M(S_h)$ and $\mathfrak{M}_h = \mathfrak{M} \cap M(S_h)$ (recall that $S_h = \{s \in S : h(s) = 1\}$). We conclude that $\mu^{-1} \in \mathfrak{N}$.

§9.2. **The spectrum of a measure**

9.2.1. If ω is a positive measure in $\mathfrak{M}$ for which $\omega^2 << \omega$, then it follows that $\mu * \nu << \omega$ whenever $\mu << \omega$ and $\nu << \omega$. In fact, if E is a set of ω-measure zero and $E' = \{s \times t \in S \times S : st \in E\}$ then

$$\omega \times \omega(E') = \omega^2(E) = 0$$

and hence

$$\mu * \nu(E) = (\mu \times \nu)(E') = 0.$$

It follows that $L(\omega) = \{\mu \in \mathfrak{M} : \mu << \omega\}$ is an L-subalgebra of $\mathfrak{M}$ in this case.

Subalgebras of the above form are somewhat easier to work with than general L-subalgebras for the following reason: the dual space $L(\omega)^*$ of $L(\omega)$ may be identified with $L^\infty(\omega)$. This means, in particular, that each complex homomorphism of $L(\omega)$ is determined by a function in $L^\infty(\omega)$. Now typically, given a measure algebra $\mathfrak{M}$, we have no concrete description of S, $\hat{S}$, or the representation of $\mathfrak{M}$ in $M(S)$. However, for a given $\omega \in \mathfrak{M}$, we will ordinarily have a quite explicit description of $L^\infty(\omega)$.

The complex homomorphisms of an algebra $L(\omega)$ are easily described.

9.2.2. Proposition. *If ω is a positive measure with $\omega^2 << \omega$, then each complex homomorphism of $L(\omega)$ has the form $\mu \to \int f\, d\mu$, where f is a nonzero element of $L^\infty(\omega)$ such that $f(st) = f(s)f(t)$ a. e. $/\omega \times \omega$.*

Proof. An element $f \in L^\infty(\omega)$ determines a complex homomorphism if and only if

$$\int_{S\times S} f(st)\, d(\mu \times \nu)(s, t) = \int_S f\, d\mu \int_S f\, d\nu = \int_{S\times S} f(s)f(t)\, d(\mu * \nu)(s, t)$$

for every pair $\mu, \nu \in L(\omega)$. This happens if and only if

$$\int_{S\times S}(f(st)-f(s)f(t))\,d\rho(s,t)=0$$

for every $\rho \ll \omega \times \omega$, that is, if and only if $f(st) = f(s)f(t)$ a.e./$\omega \times \omega$.

Note that the above condition on an element of $L^\infty(\omega)$ is independent of how $\mathfrak{M}$ is represented as an algebra of measures on a semigroup. That is, we need not be working with the representation of $\mathfrak{M}$ on its structure semigroup.

We shall call an element of $L^\infty(\omega)$ which satisfies the above condition essentially multiplicative. The set of nonzero essentially multiplicative functions in $L^\infty(\omega)$ will be denoted $\Delta(\omega)$.

We now have the following corollary of Theorem 9.1.2.

9.2.3. COROLLARY. *Let ω be a positive measure in $\mathfrak{M}$ such that $\omega^2 \ll \omega$. Furthermore, if ω_0 is the part of ω lying in the spine $\mathfrak{M}_0$ of $\mathfrak{M}$, assume that ω_0 and $\tilde{\omega}_0$ are mutually absolutely continuous. Then for each $\mu \in L(\omega)$, the spectrum of μ, as an element of $\mathfrak{M}$, is $\{\hat{\mu}(f) = \int f\,d\mu : f \in \Delta(\omega)\}$.*

PROOF. The condition that ω_0 and $\tilde{\omega}_0$ are mutually absolutely continuous implies that $L(\omega) \cap L_h$ is symmetric in L_h and hence is a group algebra for each maximal group algebra $L_h \subset \mathfrak{M}_0$. Hence $L(\omega)$ is balanced. It remains balanced if we adjoin the identity. That is, $L(\omega) + \mathbf{C}\delta$ is also balanced.

If $\lambda \in \mathbf{C}, \mu \in L(\omega)$, then by Theorem 9.1.2 we have that $\lambda\delta - \mu$ is invertible in $\mathfrak{M}$ if and only if it is invertible in $L(\omega) + \mathbf{C}\delta$. Since the spectrum of $L(\omega)$ is $\Delta(\omega)$, the corollary follows.

9.2.4. CONSTRUCTING A BALANCED ω. We now show that each $\mu \in \mathfrak{M}$ is in $L(\omega)$ for some measure ω satisfying the conditions of 9.2.3.

If $\mu \in \mathfrak{M}$ and $|\mu|_0$ is the part of $|\mu|$ in $\mathfrak{M}_0$ (that is, $|\mu|_0 \in \mathfrak{M}_0$ and $|\mu| - |\mu|_0 \in \mathfrak{M}_0^\perp$), then we set $\nu_1 = |\mu| + |\mu|_0^\sim$. Note that $(\nu_1)_0^\sim = (\nu_1)_0$; that is, the part of ν_1 which lies in $\mathfrak{M}_0$ is symmetric. We call a measure with this property balanced.

Now suppose we have defined positive balanced measures $\nu_1, \nu_2, \cdots, \nu_{n-1}$ in such a way that $\nu_i * \nu_j \ll \nu_{i+j}$ if $i, j \geqslant 1$ and $i + j < n$. We then set

$$\nu_n' = \sum_{i+j=n} \nu_i * \nu_j \quad \text{and} \quad \nu_n = \nu_n' + (\nu_n')_0^\sim.$$

It follows that ν_n is positive, balanced, and satsifies $\nu_i * \nu_j \ll \nu_n$ for $i + j = n$. By induction, there is an infinite sequence $\{\nu_n\}_{n=1}^\infty$ of balanced, positive measures such that $\mu \ll \nu_1$ and $\nu_i * \nu_j \ll \nu_{i+j}$ for $i, j \geqslant 1$.

If we set $\omega = \Sigma_{n=1}^\infty 2^{-n} \|\nu_n\|^{-1}\nu_n$, then ω is positive, balanced, and satisfies $\omega^2 \ll \omega$. Thus,

PROPOSITION. *For each measure $\mu \in \mathfrak{M}$, there exists a positive, balanced measure $\omega \in \mathfrak{M}$ such that $\mu \in L(\omega)$ and $\omega^2 \ll \omega$.*

9.2.5. MEASURES WITH SINGULAR POWERS. As an application of Corollary 9.2.3, we compute the spectrum of a positive measure $\mu \in \mathfrak{M}$ such that $\mu^n \perp \mu^m$ for $n \neq m$ and $\mu^n \perp \mathfrak{M}_0$ for all $n > 0$.

We set $\omega = \Sigma 2^{-n} \|\mu\|^{-n}\mu^n$. Note that $\omega^2 << \omega$ and, trivially, ω is balanced.

Choose disjoint Borel sets $E_1, E_2, \cdots$ such that μ^n is concentrated on E_n, but $\mu^m(E_n) = 0$ for $m \neq n$. We can do this since $\mu^n \perp \mu^m$ for $n \neq m$. For each $\lambda \in \mathbf{C}$, with $|\lambda| \leqslant 1$, we define a function $f_\lambda \in L^\infty(\omega)$ by

$$f_\lambda = \lambda^n \quad \text{on} \quad E_n, \quad \text{for} \quad n = 1, 2, \cdots.$$

Observe that $f_\lambda = \lambda^n$ a.e. $/\mu^n$ for each n. Hence,

$$f_\lambda(st) = \lambda^{i+j} = f_\lambda(s) f_\lambda(t) \qquad \text{a.e.} /\mu_i \times \mu_j.$$

It follows that $f_\lambda \in \Delta(\omega)$ and hence $\lambda \|\mu\| = \int f_\lambda \, d\mu$ is an element of the spectrum of μ. Thus,

PROPOSITION. *If* $0 \leqslant \mu \in \mathfrak{M}$, $\mu^n \perp \mathfrak{M}_0$ *for* $n \geqslant 0$, *and* $\mu^n \perp \mu^m$ *for* $n \neq m$, *then the spectrum of* μ *in* $\mathfrak{M}$ *is the disc of radius* $\|\mu\|$.

This is a simple-looking result. However, we have not been able to prove it by elementary means, even in the case $\mathfrak{M} = M(R)$. The above proof depends heavily on Theorem 9.1.2, which in turn depends on all the machinery of Chapters 6 and 7.

§9.3. The case $\mathfrak{M} = M(G)$

9.3.1. The involution in the algebra $M(G)$ makes it easy to obtain balanced measures. In fact, the involution $\mu \to \tilde{\mu}$ on $M(G)$ agrees on the spine of $M(G)$ with the involution of 7.6.7. Hence, if $\mu = \tilde{\mu} \in M(G)$, then μ is balanced.

Given a measure $\mu \in M(G)$, we set $\nu = |\mu| + |\mu|^{\sim}$ and $\omega = \Sigma 2^{-n} \|\nu\|^{-n}\nu^n$. Then $\omega^2 << \omega$, ω is balanced, and $\mu \in L(\omega)$. Thus by 9.2.3 we have

PROPOSITION. *If* $\mu \in M(G)$ *and* ω *is the measure constructed above for* μ, *then the spectrum of* μ *in* $M(G)$ *is* $\{\int f \, d\mu : f \in \Delta(\omega)\}$.

9.3.2. CONTINUOUS MEASURES IN $M(R)$. The situation for $M(R)$ is particularly simple due to the fact that the spine of $M(R)$ is just $L(R) + M_d(R)$.

For a continuous measure $\mu \in M(R)$, let $\omega = \Sigma_{n=1}^\infty 2^{-n} \|\mu\|^{-n} |\mu|^n$. The part of ω which lies in $L(R) + M_d(R)$ is just the absolutely continuous part of ω. Hence we obtain a balanced L-subalgebra $\mathfrak{N}$ of $M(R)$ by setting $\mathfrak{N} = L(\omega) + L(R)$.

Now an element of the spectrum of $\mathfrak{N}$ which fails to vanish on $L(R)$ must be given by a character of G. On the other hand, an element of the spectrum of $\mathfrak{N}$ which does vanish on $L(R)$ will be given by an element of $\Delta(\omega)$ which is zero a.e. with respect to the absolutely continuous part of ω. Thus,

PROPOSITION. *Let* μ *be a continuous measure in* $M(R)$, $\omega = \Sigma_{n=1}^\infty 2^{-n} \|\mu\|^{-n} |\mu|$, *and* ω_0 *the absolutely continuous part of* ω. *Then the spectrum of* μ *in* $M(G)$ *consists of the numbers of the form* $\int f \, d\mu$, *where either*

(i) $f \in \hat{G}$; *or*

(ii) $f \in \Delta(\omega)$ *and* $f = 0$ *a.e.* $/\omega_0$.

9.3.3. MEASURES WITH SINGULAR POWERS IN $M(R)$. If $\mu \in M(R)$ is continuous, positive, and satisfies $\mu^n \perp \mu^m$ for $n \neq m$, then we claim that $\mu^n \perp L(R)$ for all $n \geqslant 1$.

Suppose that, for some n, μ^n has a nonzero absolutely continuous part. Then we can choose a measure $\nu \neq 0$ of the form $\nu = fm$ (m is Lebesgue measure) with f bounded, positive, and compactly supported and $\nu << \mu^n$. It follows that $\nu^2 = (f * f)m$ and $f * f$ is continuous. Now if the powers of μ are mutually singular, then it follows that the powers of ν are also. However, this implies that the open set $U = \{x : (f*f)(x) > 0\}$ has the property that $U^i \cap U^j = \varnothing$ for $i \neq j$. This is impossible. Hence, we conclude that $\mu^n \perp L(R)$ for all n if $\mu^n \perp \mu^m$ for $n \neq m$.

We now have that a positive continuous measure $\mu \in M(R)$ with singular powers has the property that $\mu^n \perp L(R) + M_d(R)$ for all n. Hence, by 9.2.5, we have:

PROPOSITION. *If $\mu \in M(R)$ is positive, continuous, and satisfies $\mu^n \perp \mu^m$ for $n \neq m$, then the spectrum of μ is the disc of radius $\|\mu\|$.*

CHAPTER 10. BOUNDARIES AND GLEASON PARTS

It turns out that the semigroup structure of the spectrum $\hat{S}$ of a measure algebra can be used to give partial descriptions of the Shilov boundary, the strong boundary, and the Gleason parts for $\mathfrak{M}$. The results here are still incomplete and several problems remain to be solved.

In this chapter we present partial results on the Shilov and strong boundaries and discuss analytic structure and Gleason parts. We give a complete characterization of the one-point parts. For a nondiscrete l.c.a. group G, we discuss the result of Taylor [2] and Johnson [3] that the Shilov boundary is proper in $\hat{S}$. Also, we present some recent results of Brown and Moran [3] on infinite product measures.

§10.1. Analytic structure in S

10.1.1. Let A be a commutative Banach algebra with spectrum Δ and let $U \subset \mathbf{C}$ be a domain. If $\alpha : U \to \Delta$ is a map such that $\hat{a} \circ \alpha : U \to \mathbf{C}$ is analytic for each $a \in A$, then we shall say that α is analytic.

Proposition. *Let $\mathfrak{M}$ be a commutative, semisimple measure algebra with structure semigroup S. If $f \in \hat{S}$ and $g \in \hat{S}^+$, then $z \to fg^z : \{z \in \mathbf{C} : \mathrm{Re}\,(z) > 0\} \to \hat{S}$ is an analytic map. This map is constant if and only if $g^2 = g$.*

Proof. For each $\mu \in \mathfrak{M}$, the map $z \to \hat{\mu}(fg^z) = \int fg^z \, d\mu$ is clearly analytic for $\mathrm{Re}\,(z) > 0$.

10.1.2. Boundaries. If A is a commutative Banach algebra with spectrum Δ, then $h \in \Delta$ is called a strong boundary point if for each neighborhood U of h, there is an element $a \in A$ such that $|\hat{a}|$ achieves its maximum at h but not at any point of $\Delta \backslash U$.

A boundary for A is a set $\Lambda \subset \Delta$ such that $|\hat{a}|$ achieves its maximum on Λ for each $a \in A$. The set of all strong boundary points is a boundary for A and is called the strong boundary (Gamelin [1]). There is a unique minimal closed boundary for A. This is the Shilov boundary (Gamelin [1]). The strong boundary is a dense subset of the Shilov boundary.

Suppose $h \in \Delta$, z is a point of a connected domain $U \subset \mathbf{C}$, and $\alpha : U \to \Delta$ is a nonconstant analytic map with $\alpha(z) = h$. Then it follows from the maximum modulus principle that h cannot be a strong boundary point for A.

Thus by 10.1.1 we have:

PROPOSITION. *If* $\mathfrak{M}$ *is a commutative, semisimple measure algebra with structure semigroup* S, *then each strong boundary point* $f \in \hat{S}$ *satisfies* $|f|^2 = |f|$. *That is,* f *is in some maximal group in* $\hat{S}$.

PROOF. If $|f|^2 \neq |f|$, let $f = h|f|$ be the polar decomposition of f (3.4.4). We then have that $z \to h|f|^z$ (Re $(z) > 0$) is a nonconstant analytic map having the value f at 1. Hence f is not a strong boundary point.

10.1.3. GLEASON PARTS. Let A be a commutative Banach algebra with spectrum Δ. The completion of $\hat{A} = \{\hat{a} : a \in A\}$ in sup norm on Δ is a uniform algebra with spectrum Δ. For such an algebra, Gleason [1] pointed out that the relation $\sim$ on Δ, defined by $h \sim k$ if and only if $\|h - k\| < 2$, is an equivalence relation. Here, $\|h - k\|$ denotes the norm of $h - k$ in the adjoint space of the uniform algebra. In terms of our original algebra A this is given by

$$\|h - k\| = \sup \{|\hat{a}(h) - \hat{a}(k)| : a \in A, \|\hat{a}\|_\infty \leqslant 1\}.$$

An equivalence class in Δ, under this relation, is called a part.

There are several ways of defining the relation $\sim$. We state some of these without proof (cf. Gamelin [1]).

PROPOSITION. *For a commutative Banach algebra* A *with spectrum* Δ *and for* h, $k \in \Delta$, *the following are equivalent*:

(a) h *and* k *are in the same part*;

(b) $\sup \{|\hat{a}(h)| : a \in A, \hat{a}(k) = 0, \|\hat{a}\|_\infty \leqslant 1\} < 1$;

(c) *there is a positive number* c *such that* $c^{-1} \leqslant u(h)u(k)^{-1} \leqslant c$ *whenever* $u = \text{Re}(\hat{a}) > 0$ *for some* $a \in A$;

(d) *if* $\{a_\alpha\}$ *is a net in* A *with* $\|\hat{a}_\alpha\| \leqslant 1$ *for all* α *and if* $\hat{a}_\alpha(h) \to 1$, *then also* $\hat{a}_\alpha(k) \to 1$.

A representing measure for $h \in \Delta$ is a positive, normalized measure $\mu \in M(\Delta)$ such that

$$\hat{a}(h) = \int \hat{a}(k)\, d\mu(k)$$

for each $a \in A$. It follows from part (c) of the above proposition that if h and k are in the same part then there are representing measures μ for h and ν for k such that μ and ν are mutually absolutely continuous and the Radon-Nikodym derivative $d\mu/d\nu$ satisfies $c^{-1} \leqslant d\mu/d\nu \leqslant c$ (Gamelin [1]). On the other hand, if h and k are not in the same part, then it follows from the definition that any pair of representing measures for h and k must be mutually singular (Gamelin [1]).

10.1.4. ANALYTIC STRUCTURE AND PARTS. It follows from Harnak's inequality and (c) of 10.1.3 that the image of an analytic map, $\alpha : U \to \Delta$ (U a connected domain in **C**) must lie in a single part of Δ. Thus,

PROPOSITION. *Let* $\mathfrak{M}$ *be a measure algebra with structure semigroup* S. *Then for each* $f \in \hat{S}$, $g \in \hat{S}^+$, *the set* $\{fg^z : \text{Re}(z) > 0\}$ *is contained in a single part of* $\hat{S}$.

§10.2. Miller's theorem on parts

10.2.1. In this section $\mathfrak{M}$ will denote a commutative, semisimple measure algebra with structure semigroup S.

On $\hat{S}$ we introduce an equivalence relation due to Miller [1].

DEFINITION. If $f, g \in \hat{S}$ then we will write $f \approx g$, provided f and g are equal on the set $\{s \in S : |f(s)| = 1\} \cup \{s \in S : |g(s)| = 1\}$.

Note that the sets $\{s \in S : |f(s)| = 1\}$ and $\{s \in S : |g(s)| = 1\}$ are identical if $f \approx g$. Obviously $\approx$ is an equivalence relation. Miller conjectures that the relation $\approx$ agrees with the relation $\sim$ defining parts. He proves this in one direction and deals with a special case of the other direction. We outline his results here.

10.2.2. PROPOSITION. *If $f, g \in \hat{S}$ and $f \sim g$, then $f \approx g$.*

PROOF. Suppose it is not true that $f \approx g$. Then, without loss of generality, we may assume that f and g differ at some point, say s, for which $|f(s)| = 1$. Now let $\{\mu_\alpha\}$ be a net of normalized positive measures in $\mathfrak{M}$ such that $\mu_\alpha \to \delta_s$ is the weak-$*$ topology of $M(S)$. If we set $c = \overline{f(s)}$, then $\|(c\mu_\alpha)^\wedge\|_\infty \leqslant 1$, $(c\mu_\alpha)^\wedge(f) \to cf(s) = 1$, and $(c\mu_\alpha)^\wedge(g) \to cg(s) \neq 1$. Hence, by 10.1.3, f and g are not in the same part.

10.2.3. LEMMA. *Let f, g be in $\hat{S}$ and have polar decompositions $f = h|f|$ and $g = k|g|$. If there exists $r < 1$ such that $f(s) = g(s)$ whenever $|f(s)| > r$ or $|g(s)| > r$, then*

(i) $$|h|f|^z - k|g|^z| \leqslant 2r^{\operatorname{Re}(z)}$$

for all $z \in \mathbf{C}$ with $\operatorname{Re}(z) > 0$.

PROOF. If $s \in S$ is such that $|f(s)| \leqslant r$ and $|g(s)| \leqslant r$ then the inequality obviously holds. However, for other values of s we have $f(s) = g(s)$ and hence $h(s) = k(s)$. The left side of (i) is zero for these values of s.

10.2.4. PROPOSITION. *If $f, g \in \hat{S}$ and there exists $r < 1$ such that $f(s) = g(s)$ whenever $|f(s)| > r$ or $|g(s)| > r$, then f and g are in the same part.*

PROOF. Let $f = h|f|$ and $g = k|f|$ be the polar decompositions of f and g, and for $\mu \in \mathfrak{M}$ consider the analytic function ϕ on $\{z \in \mathbf{C} : \operatorname{Re}(z) > 0\}$,

$$\phi(z) = \hat{\mu}(h|f|^z) - \hat{\mu}(k|g|^z) = \int (h|f|^z - k|g|^z)\, d\mu.$$

By the previous lemma we have

$$|\phi(z)| \leqslant 2\|\mu\| r^{\operatorname{Re}(z)}.$$

It follows that $\phi(z)r^{-z}$ is bounded and analytic on the right half plane.

If we choose μ such that $\|\hat{\mu}\|_\infty \leqslant 1$, then we have $|\phi(z)| \leqslant 2$. Since $|r^{-z}| \to 1$ as $\operatorname{Re}(z) \to 0$, it follows from the maximum modulus principle that $|\phi(z)r^{-z}| \leqslant 2$ as well. We conclude that

$$|\hat{\mu}(f) - \hat{\mu}(g)| = |\phi(1)| \leqslant 2r < 2.$$

Since r is independent of μ, we conclude that f and g are in the same part.

The multiplication in $\hat{S}$ respects parts. In fact,

10.2.5. LEMMA. *If $f, g, h \in \hat{S}$ and $f \sim g$, then $fh \sim gh$.*

PROOF. If $f \sim g$ then for some $r < 2$ we have $|\hat{\mu}(f) - \hat{\mu}(g)| \leqslant r$ for all $\mu \in \mathfrak{M}$ with $\|\hat{\mu}\|_\infty \leqslant 1$. However, if $\|\hat{\mu}\|_\infty \leqslant 1$ then also $\|(h\mu)\,\hat{}\,\|_\infty = \sup\{|\hat{\mu}(hk)| : k \in \hat{S}\} \leqslant 1$. Hence

$$|\hat{\mu}(hf) - \hat{\mu}(hg)| = |(h\mu)\,\hat{}\,(f) - (h\mu)\,\hat{}\,(g)| \leqslant r,$$

and $hf \sim hg$.

Note that this lemma implies that the product of two parts in $\hat{S}$ is contained in a part. Hence $\hat{S}/(\sim)$ inherits a semigroup structure from $\hat{S}$.

10.2.6. ONE-POINT PARTS. Suppose $f \in \hat{S}$ has the property that $\{f\}$ is a part. It follows from 10.1.4 that $|f|^2 = |f|$. That is, f is in the maximal group in $\hat{S}$ with idempotent $|f|$. We claim that Miller's conjecture holds in this case; that is, there is no $g \in \hat{S}$, with $g \neq f$, such that $g \approx f$.

If $g \in \hat{S}$ satisfies $g \approx f$, then $U = \{s \in S : |g(s)| = 1\} = \{s \in S : |f(s)| = 1\}$ and $g(s) = f(s)$ on U. Since $|f|^2 = |f|$, U is open and closed. Since $|g(s)| < 1$ for s in the compact set $S \backslash U$, we have $|g(s)| \leqslant r$ on $S \backslash U$ for some $r < 1$. However, by 10.2.4, this implies that $f \sim g$. Hence, if $\{f\}$ is to be a one-point part we must have $f = g$.

Since $f \sim g$ implies $f \approx g$, by 10.2.1, we have:

THEOREM. *If $f \in \hat{S}$ then $\{f\}$ is a part if and only if $f \approx g \in \hat{S}$ implies $f = g$. Furthermore, if $\{f\}$ is a part then $|f|^2 = |f|$.*

Miller's main theorem is the following.

10.2.7. THEOREM. *For $f, g \in \hat{S}^+$, $f \sim g$ if and only if $f \approx g$, that is, if and only if $\{s \in S : f(s) = 1\} = \{s \in S : g(s) = 1\}$.*

We give only an outline of the proof:

Suppose $f \approx g$ and let $h = (fg)^0$. Note that $h = 1$ in a neighborhood of $\{s \in S : f(s) = 1\} = \{s \in S : g(s) = 1\}$. Furthermore, on $S_h = \{s \in S : h(s) = 1\}$ each of f and g is strictly positive a.e./μ for each $\mu \in \mathfrak{M}$.

Now it follows from 10.2.4 that $f \sim fh$ and $g \sim gh$. Thus, in order to prove that $f \sim g$ it suffices to prove that $fh \sim gh$. Also, observe that for $\mu \in \mathfrak{M}$, $\hat{\mu}(fh) = (h\mu)\,\hat{}\,(fh)$, $\hat{\mu}(gh) = (h\mu)\,\hat{}\,(gh)$, and $h\mu \in \mathfrak{M}_h = \mathfrak{M} \cap M(S_h)$. It follows that we may as well consider fh and gh to be points of the spectrum of $\mathfrak{M}_h$. If they are in the same part relative to this algebra they will be in the same part relative to $\mathfrak{M}$.

We conclude from the above that it suffices to prove the theorem in the special case where $f, g > 0$ a.e./μ for each $\mu \in \mathfrak{M}$. Under this assumption, $\mathfrak{M}$ is supported on the subsemigroup $S_1 = \{s \in S : f(s) > 0,\ g(s) > 0\}$.

We define a homomorphism $\alpha: S_1 \to R^+ \times R^+$ by $\alpha(s) = (-\ln f(s), -\ln g(s))$. If $T = \alpha(S_1) \subset R^+ \times R^+$, then $\mu \to \mu \circ \alpha^{-1}$ is a homomorphism of measure algebras from $\mathfrak{M}$ into $M(T)$. The semicharacters $(x, y) \to e^{-x}$ and $(x, y) \to e^{-y}$ on T determine complex homomorphisms of $M(T)$ which map to f and g, respectively, under the induced map from the spectrum of $M(T)$ to the spectrum of $\mathfrak{M}$.

Now it is easy to see that if $\phi: A \to B$ is a Banach algebra homomorphism, then the induced map $\phi^*: \Delta(B) \to \Delta(A)$ maps parts into parts. Hence, it follows from the above that $f \sim g$ provided $(x, y) \to e^{-x}$ and $(x, y) \to e^{-y}$ determine equivalent points of the spectrum of $M(T)$.

The key to the remainder of the proof is the geometry of the set T. The hypothesis that $\{s \in S: f(s) = 1\} = \{s \in S: g(s) = 1\}$ implies that there are positive numbers a and k such that T is contained in the union of the wedge $V = \{(x, y) \in R^+ \times R^+: k^{-1} \leqslant y/x \leqslant k\}$ and the set $U = \{(x, y) \in R^+ \times R^+: x \geqslant a, y \geqslant a\}$.

Now let Λ be the function on R which is $a - |x|$ if $|x| \leqslant a$ and is zero otherwise. We define a function h on R^2 by

$$h(x, y) = e^{-|x|} - \epsilon e^{-k|x|}\Lambda(x) + \epsilon e^{-|kx-y|}e^{-|y|}\Lambda(x).$$

Note that $h(0, 0) = 1$. For sufficiently small ϵ, Miller proves that this function is positive definite and, hence, is the Fourier transform of a positive normalized measure $\nu_1 \in M(R^2)$. Furthermore, $h(x, y) = e^{-x}$ on T. Hence, for $\mu \in M(T)$ we have

$$\int\left[\int e^{-i(ux+vy)}\, d\mu(x, y)\right] d\nu_1(u, v) = \int \hat{\nu}_1(x, y)\, d\mu(x, y) = \int e^{-x}\, d\mu(x, y).$$

In other words, ν_1 is a representing measure for the complex homomorphism of $M(T)$ determined by e^{-x}, with support on the set of complex homomorphisms determined by characters of R^2.

Now by interchanging the roles of x and y in the above, we obtain a representing measure ν_2 for e^{-y}. However, ν_1 and ν_2 are not mutually singular due to the third term in the expression for h. In fact, this term is the transform of an absolutely continuous measure with an analytic Radon-Nikodym derivative, while the other two terms are transforms of singular measures. The theorem now follows from 10.1.3.

10.2.8. REMARK. In view of the preceding result, Miller's conjecture reduces to the following problem: If $f \in \hat{S}$ and $|f(s)| = 1$ implies $f(s) = 1$, then is $f \sim |f|$?

The reduction proceeds as follows: let f and g be elements of $\hat{S}$ with $f \approx g$. Then it is also true that $f \approx f|g|$ and $g \approx f|g|$. Furthermore, $|f|\,|g| \leqslant |f|$ and $|f|\,|g| \leqslant |g|$. It follows that to prove that $f \approx g$ implies $f \sim g$ it suffices to consider the case where $|g| \leqslant |f|$. With this assumption, let $f = h|f|$ be the polar decomposition of f and note that $g = h\bar{h}g$ and $|\bar{h}g|\,(s) = 1$ implies that $\bar{h}g(s) = 1$. Furthermore, $|f| \approx |g|$ and so $|f| \sim |g|$ by 10.2.7. If we knew that $\bar{h}g \sim |\bar{h}g| = |g|$ we could conclude that $f = h|f| \sim h|g| \sim h\bar{h}g = g$.

10.2.9. EXAMPLES. Consider the algebra $M(Z^+)$, where $Z^+ = \{n \in Z: n \geqslant 0\}$. Here the structure semigroup S is the Bohr compactification of Z^+ and $\hat{S} = \hat{Z}^+$ is the unit disc in $\mathbf{C}$. For $f \in \hat{S}$ the set $\{s \in S : |f(s)| = 1\}$ is either $\{0\}$ (for $f(n) = z^n$ with z in the interior of the disc) or it is all of S (for $f(n) = z^n$ with z on the boundary of the disc). It follows from 10.2.6 that the points on the boundary of the disc are one-point parts. By 10.2.4 the interior of the disc is a single part.

Now consider the algebra $L(R^+) + \mathbf{C}\delta$. Here S is the discrete union of the Bohr compactification S_0 of R^+ with a singleton set $\{e\}$, where e acts as an identity for S. There is a semicharacter $f_\infty \in \hat{S}$ defined by $f_\infty(e) = 1$ and $f_\infty(s) = 0$ for $s \neq e$. All the other semicharacters in $\hat{S}$ are extensions to S of the functions f_z $(z \in \mathbf{C} : \operatorname{Re}(z) \geqslant 0)$ defined by $f_z(e) = 1, f_z(x) = e^{-zx}$ for $x \in R^+$.

Note that f_∞ is the only element $f \in \hat{S}$ for which $\{s \in S : |f(s)| = 1\} = \{e\}$. Hence, $\{f_\infty\}$ is a one-point part. For $\operatorname{Re}(z) = 0$, $\{s \in S : |f_z(s)| = 1\} = S$ and, hence, $\{f_z\}$ is a one-point part. The set $\{f_z : \operatorname{Re}(z) > 0\}$ is a single part by 10.1.4.

An interesting thing happens if we modify the above example by restricting to the subalgebra $L([1, \infty)) + \mathbf{C}\delta \subset L(R^+) + \mathbf{C}\delta$. Here the structure semigroup is $\{e\}$ union the Bohr compactification of $[1, \infty)$. This is a subsemigroup S_1 of the semigroup S above. However, note that $\hat{S}_1 = \hat{S}$. Thus the spectrum of $L([1, \infty)) + \mathbf{C}\delta$ is the same as the spectrum of $L(R^+) + \mathbf{C}\delta$. One thing about the spectrum does change, however. The point f_∞ is now in the same part with $\{f_z : \operatorname{Re}(z) > 0\}$. In fact, on S_1, $\{s : |f_z(s)| = 1\} = \{e\} = \{s : f_z(s) = 1\}$ if $\operatorname{Re}(z) > 0$. It follows from 10.2.4 that $\{f_\infty\} \cup \{f_z : \operatorname{Re}(z) > 0\}$ is a part.

§10.3. The strong boundary

10.3.1. If A is a commutative Banach algebra with spectrum Δ and if $h \in \Delta$ is a strong boundary point, then h has a unique representing measure δ_h. In fact, if $\mu \in M(\Delta)$ is a representing measure for h, U a neighborhood of h, and $a \in A$ is such that $|\hat{a}(h)| = \|\hat{a}\|_\infty$ and $|\hat{a}| < \|\hat{a}\|_\infty$ on $\Delta \backslash U$, then

$$\|\hat{a}\|_\infty = |\hat{a}(h)| = \left| \int_U \hat{a}(k)\, d\mu(k) + \int_{\Delta\backslash U} \hat{a}(k)\, d\mu(k) \right|$$

$$\leqslant \|\hat{a}\|_\infty \mu(U) + \sup_{k \in \Delta\backslash U} |\hat{a}(k)|\, \mu(\Delta\backslash U).$$

Since $\mu(U) + \mu(\Delta\backslash U) = \mu(\Delta) = 1$, we conclude that μ must be supported on U. Since this is true for every neighborhood U of h, we conclude that $\mu = \delta_h$.

Now, by 10.1.3, a point of Δ with a unique representing measure must be a one-point part. Hence, each strong boundary point of Δ is a one-point part. We conclude from 10.2.6 that:

10.3.2. PROPOSITION. *If $f \in \hat{S}$ is a strong boundary point for the measure algebra $\mathfrak{M}$, then $|f|^2 = |f|$ and there is no $g \in \hat{S}$, distinct from f, such that $f(s) = g(s)$ whenever $|f(s)| = 1$ or $|g(s)| = 1$.*

There is a class of examples, due to Hoffman [1], of algebras with one-point parts not in the Shilov boundary. One of these is used by Garnett [1] to construct algebras with spectra containing parts which have a quite general topological structure. It turns out that the example used by Garnett is the completion, in spectral norm, of a certain measure algebra. We describe this measure algebra below. Its existence shows that the converse of the above proposition is false. This leaves open the problem of completely determining the strong boundary for a measure algebra.

10.3.3. AN EXAMPLE. Let $r > 0$ be an irrational number and let $T = \{(n, m) \in Z \times Z : rn + m \geqslant 0\}$. Note that T is a subsemigroup of $Z \times Z$ containing the identity $(0, 0)$. By 4.1.3 the structure semigroup of $M(T)$ is $\overline{T}$, the Bohr compactification of T, while the spectrum of $M(T)$ is $\hat{T}$.

Let $f_\infty \in \hat{T}$ denote the semicharacter such that $f_\infty(0, 0) = 1$ and $f_\infty(n, m) = 0$ if $n \neq 0$ or $m \neq 0$. The element f_∞ is not in the Shilov boundary for $M(T)$. In fact, by the results of Arens-Singer (4.4.2), the Shilov boundary for $M(T)$ is the restriction of $(Z \times Z)^\wedge$ to T. We shall show that $\{f_\infty\}$ is a one-point part.

10.3.4. *For the algebra $M(T)$ above the singleton set $\{f_\infty\} \subset \hat{T}$ is a part.*

PROOF. We first prove that if $(0, 0) \neq t \in T$ then $t = t_1 + t_2$ for a pair $t_1, t_2 \in T$ with $t_1 \neq (0, 0) \neq t_2$.

Let $t = (n, m)$. Note that $t \neq (0, 0)$ implies that $rn + m > 0$ since r is irrational. Now we can clearly choose integers n_1 and m_1 such that $rn + m > rn_1 + m_1 > 0$. If we set $n_2 = n - n_1$ and $m_2 = m - m_1$, then the elements $t_1 = (n_1, m_1)$ and $t_2 = (n_2, m_2)$ are in T and $t = t_1 + t_2$.

Now suppose $f \in \hat{T}$ and $f \neq f_\infty$. There exists $t \neq (0, 0)$ such that $f(t) \neq 0$. Furthermore if $t = t_1 + t_2$, as above, then $f(t) = f(t_1)f(t_2)$. It follows that $|f(t_i)| \geqslant |f(t)|^{1/2}$ for either $i = 1$ or $i = 2$. By iterating this procedure, we conclude that there is a sequence of points of T, distinct from $(0, 0)$, on which $|f|$ approaches 1. If $s \in \overline{T}$ is a limit point of such a sequence, then $|f(s)| = 1$ and $s \neq (0, 0)$. It follows from 10.2.2 that f and f_∞ are not in the same part. Since f was an arbitrary element of $\hat{T} \backslash \{f_\infty\}$, we conclude that $\{f_\infty\}$ is a one-point part.

§10.4. Infinite convolution products

10.4.1. Our discussion of the Shilov boundary of $M(G)$ in §10.5 will require the construction of very special kinds of measures on G. There is a class of measures which are surprisingly easy to deal with in this regard–the infinite convolution products of discrete measures. Williamson [1] gives the following criteria for the convergence of infinite product measures.

PROPOSITION. *Let $\{V_i\}_{i=0}^\infty$ be a neighborhood base at $e \in G$ such that each V_i is symmetric and has compact closure and $V_i \cdot V_{i+1} \subset V_i$ for each i. If, for each i, μ_i is a positive normalized measure concentrated on V_i, then the infinite product $\Pi_{i=1}^\infty \mu_i$ converges in the weak-$*$ topology to a measure concentrated on V_0.*

PROOF. Let $\nu_n = \Pi_{i=1}^n \mu_i$ and $\omega_{nm} = \Pi_{i=n+1}^m \mu_i$. Note that ω_{nm} is concentrated on V_n. Since V_n is symmetric, $\widetilde{\omega}_{nm}$ is also concentrated on V_n. It follows that $\widetilde{\omega}_{nm} * f \to f$ uniformly as $n, m \to \infty$ for each $f \in C_0(G)$. Furthermore,

$$\int f\,dvn - \int f\,dvm = \int (f - \widetilde{\omega}_{nm} * f)\,dv_n.$$

It follows that $\{\int f\,d\mu\}$ is a Cauchy sequence for each $f \in C_0(G)$ and, hence, $\{\nu_n\}$ converges in the weak-$*$ topology of $M(G)$.

10.4.2. Brown and Moran [3] have observed that infinite products of discrete measures possess very special properties. We shall discuss some of these properties here.

In general, when we say that μ is the infinite product $\Pi_{i=1}^{\infty} \mu_i$ of the measures μ_i, we shall mean that μ is the weak-$*$ limit of the sequence $\{\Pi_{i=1}^{n} \mu_i\}$ in the dual space of the bounded continuous functions on G.

PROPOSITION. *Let* $\mathfrak{M}$ *be a translation invariant* L*-subspace of* $M(G)$. *If* $\mu = \Pi_{i=1}^{\infty} \mu_i$ *is an infinite product of normalized discrete measures* μ_i, *then either* $\mu \in \mathfrak{M}$ *or* $\mu \perp \mathfrak{M}$.

PROOF. For each n, let $\nu_n = \Pi_{i=1}^{n} \mu_i$ and $\omega_n = \Pi_{i=n+1}^{\infty} \mu_i$. Note that, for each n, $\mu = \nu_n * \omega_n$, $\|\nu_n\| \leqslant 1$, $\|\omega_n\| \leqslant 1$, and ν_n is discrete.

Let $\omega \in M(G)$ be a weak-$*$ cluster point of the sequence $\{\omega_n\}$. If $\alpha \in \hat{G}$ then either the sequence $\{\hat{\omega}_n(\alpha)\}$ is identically zero in which case $\hat{\omega}(\alpha) = 0$, or $\hat{\omega}_n(\alpha) \neq 0$ for some n. However, since $\omega_n = (\Pi_{i=n+1}^{m} \mu_i) * \omega_m$, on passing to the limit we conclude that $\omega_n = \omega_n * \omega$ and hence that $\hat{\omega}(\alpha) = 1$ if $\hat{\omega}_n(\alpha) \neq 0$. It follows that the sequence $\{\omega_n\}$ actually converges in the weak-$*$ topology and that the limit ω is an idempotent of $M(G)$. Since $\|\omega\| \leqslant 1$, we conclude that $\gamma\omega$ is the Haar measure of some compact subgroup $K \subset G$ for some $\gamma \in \hat{G}$. Without loss of generality, we may assume ω is the Haar measure of K, since we could replace μ by $\gamma\mu$ and each μ_i by $\gamma\mu_i$.

Now suppose that μ' and μ'' are the parts of μ in $\mathfrak{M}$ and $\mathfrak{M}^{\perp}$, respectively, and that ω_n' and ω_n'' are the parts of ω_n in $\mathfrak{M}$ and $\mathfrak{M}^{\perp}$. Since $\mathfrak{M}$ and $\mathfrak{M}^{\perp}$ are translation invariant and ν_n is discrete, we conclude that $\mu' = \nu_n * \omega_n'$ and $\mu'' = \nu_n * \omega_n''$. If we pass to the limit through a cofinal net of integers for which both $\{\omega_n'\}$ and $\{\omega_n''\}$ converge weak-$*$, we obtain measures ω' and ω'' such that $\mu' = \mu * \omega'$, $\mu'' = \mu * \omega''$, and $\omega = \omega' + \omega''$. Furthermore, since $\|\omega_n\| = \|\omega_n'\| + \|\omega_n''\| \leqslant 1$ and $\|\omega\| = 1$, we conclude that $\|\omega'\| + \|\omega''\| \leqslant \|\omega\|$ and hence that $\|\omega'\| + \|\omega''\| = \|\omega\|$. This implies that ω' and ω'' are positive, and absolutely continuous with respect to ω.

Now for each $g \in K$, ω'' and $\delta_g * \omega'$ are mutually singular. If not, we would have $\|\omega'' - \delta_g * \omega'\| \leqslant r < 1$ and, since $\omega_n'' \perp \delta_g * \omega_n' \in \mathfrak{M}$, $\|\omega_n''\| + \|\omega_n'\| = \|\omega_n'' - \delta_g * \omega_n'\| = \|\omega_n * (\omega'' - \delta_g * \omega')\| \leqslant r$. This is impossible since $\|\omega_n'' + \omega_n'\| = \|\omega_n\| \to 1$. Since ω'', $\omega' \in L(K)$ we conclude that $\omega'' = 0$ or $\omega' = 0$. Hence $\mu'' = 0$ or $\mu' = 0$.

10.4.3. SEMICHARACTERS AND INFINITE PRODUCTS. Let S be the structure semigroup of $M(G)$ and $\mu \to \mu_S : M(G) \to M(S)$ the canonical embedding.

Suppose $\mu = \Pi_{i=1}^{\infty} \mu_i$ is an infinite product of normalized discrete measures and $f \in \hat{S}$ is a semicharacter which is identically one on the maximal group K_e at the identity in S; that is, $\int f\,d\nu_S = \int 1\,d\nu_S$ for every discrete measure $\nu \in M(G)$. It follows that for each

$z \in C$, $\{s \in S : f(s) = z\}$ is invariant under multiplication by points of K_e and hence that for each open set $E \subset C$ the set of $\omega \in M(G)$ such that ω_S is concentrated on $f^{-1}(E)$ is a translation invariant L-subapace of $M(G)$. It follows from 10.4.2 that μ_S must be concentrated on $f^{-1}(\{c\})$ for some $c \in \mathbf{C}$ with $|c| \leqslant 1$.

If we let $f_\mu \in L^\infty(\mu)$ denote the element for which $\int f_\mu \, d\nu = \int f \, d\nu_s$ for each $\nu \in L(\mu)$, then the above result means exactly that $f_\mu = c$ a. e./μ. Thus,

PROPOSITION. *Let μ be an infinite product of normalized discrete measures and let $f \in \hat{S}$ be any complex homomorphism of $M(G)$ which agrees with 1 on the discrete measures. Then $f_\mu = c$ a. e./μ for some constant c in the unit disc.*

Note that the above hypothesis concerning f is certainly satisfied if $f \in \hat{S}^+$.

10.4.4. THE DICHOTOMY OF BROWN AND MORAN. According to the above proposition, f_μ is a constant in $[0, 1]$ if $f \in \hat{S}^+$ and μ is an infinite product of normalized discrete measures. Suppose for a given μ this constant is either zero or one for each $f \in \hat{S}^+$. Then $\{f \in \hat{S}^+ : f_\mu = 1\}$ is open and closed in $\hat{S}^+$. It follows from 5.1.6 and 5.2.4 that this set has a minimal element h which is a critical point. Furthermore, μ_S is concentrated on the kernel K_h of $S_h = h^{-1}(\{1\})$ (since $K_h = \{s \in \hat{S}^+ : h(s) = 1$ and $f(s) = 0$ for $f < h\}$).

Now by 7.1.3 and 8.2.2 there is a stronger l. c. a. group topology τ on G such that $\{\nu \in M(G) : \nu_S \in M(K_h)\} = \operatorname{Rad}(L(G_\tau))$. Hence, $\mu \in \operatorname{Rad}(L(G_\tau))$. This forces some power μ^n of μ to have a nonzero part in $L(G_\tau)$. However, by 10.4.2 this can happen only if $\mu^n \in L(G_\tau)$. Thus,

PROPOSITION. *Let μ be an infinite product of normalized discrete measures in $M(G)$. Then either $0 < f_\mu < 1$ for some $f \in \hat{S}^+$ or $\mu^n \in L(G_\tau)$ for some n and some stronger l. c. a. group topology τ on G.*

Note that if $0 < f_\mu < 1$ for some $f \in \hat{S}^+$ then, since f_μ is a constant c, for each n we have that $f_{\mu^n} = c^n$. Thus the sequence of numbers $\{f_{\mu^n}\}$ is a sequence of distinct constants. Furthermore, since $f_{\delta_g} = 1$ for each $g \in G$, we have that $f_\nu = f_{\mu^n}$ whenever ν is a translate of μ^n. Thus we have the following.

10.4.5. COROLLARY. *If μ is an infinite product of normalized discrete measures in $M(G)$ and if no power of μ is in a group algebra in $M(G)$, then for $n \neq m$ all translates of μ^n are singular with respect to μ^m.*

Another corollary of 10.4.4 is the following:

10.4.6. COROLLARY. *Let μ be an infinite product of discrete measures in $M(G)$. If no power of μ is in a group algebra, then for each c in the unit disc there is an $f \in \hat{S}$ with $f_\mu = c$.*

PROOF. Let $g \in \hat{S}^+$ be such that $g_\mu = b$ with $0 < b < 1$. Then, as z ranges over the right half plane, the numbers $b^z = g_\mu^z$ range over a dense subset of the unit disc. It follows that for each c in the unit disc there is an $f \in \hat{S}$ with $f_\mu = c$.

§10.5. The Shilov boundary of $M(G)$

10.5.1. We have shown that the strong boundary for a measure algebra is contained in $\{f \in \hat{S} : |f|^2 = |f|\}$. For $M(G)$, where G is a nondiscrete l. c. a. group, this is a proper subset of $\hat{S}$ (4.3.7). However, the Shilov boundary is the closure of the strong boundary and it is not at all easy to see that this is proper in $\hat{S}$. The fact that it is was proved by Taylor [2] for the group D_2 and by Johnson [3] in general. In each case, the key to the proof is the construction of a special kind of measure in $M(G)$.

10.5.2. PROPOSITION. *Let G be a nondiscrete l. c. a. group and let S be the structure semigroup of $M(G)$, Suppose there is a measure $\mu \in M(G)$ such that:*

(a) *for each $f \in \hat{S}$ there is a constant c with $|c| \leqslant 1$ and a $\gamma \in \hat{G}$ such that $f_\mu = c\gamma$;*

(b) *each constant c with $|c| \leqslant 1$ is f_μ for some $f \in \hat{S}$; and*

(c) *the weak-$*$ closure of $\hat{G}$ in $L^\infty(\mu)$ does not contain constants of arbitrary modulus in $(0, 1)$.*

Then the Shilov boundary of $M(G)$ is a proper subset of $\hat{S}$.

PROOF. An element f of the strong boundary satsifies $|f|^2 = |f|$. Hence, (a) implies that $f_\mu = c\gamma$ with $\gamma \in \hat{G}$ and $|c| = 1$ or $|c| = 0$. Now (c) implies that there is some $r \in (0, 1)$ such that no f in the closure of the strong boundary is such that f_μ is a constant of modulus r. However, (b) implies that $f_\mu = r$ for some $f \in \hat{S}$. Such an f cannot be in the Shilov boundary.

10.5.3. SPECIAL INFINITE PRODUCTS. Property (a) of 10.5.2 looks very strong. However, it is not difficult to achieve with certain infinite product measures.

Let $\mu = \Pi_{i=1}^\infty \mu_i$ be an infinite product of positive normalized discrete measures. As before, let $\nu_n = \Pi_{i=1}^n \mu_i$ and $\omega_n = \Pi_{i=n+1}^\infty \mu_i$. We assume that each μ_i has finite support S_i and that K_n is a Borel set supporting ω_n. We set $T_n = \Pi_{i=1}^n S_i = \text{support}(\nu_n)$ and assume that $\{gK_n : g \in T_n\}$ is a disjoint collection of sets and that each neighborhood of zero contains some K_n.

Note that $\bigcup\{gK_n : g \in T_n\}$ supports the measure μ. We let C_n be the space of simple functions on this set which are constant on gK_n for each $g \in T_n$. The fact that the K_n's are eventually in any neighborhood of the identity implies that $\bigcup_n C_n$ is dense in $L^1(\mu)$.

For $f \in \hat{S}$ and each n let $h_n \in C_n$ be chosen so as to minimize

$$\begin{aligned}\int |f_\mu - h_n|\, d\mu &= \int |f_\mu - h_n|\, d\omega_n * \nu_n \\ &= \sum_{g \in T_n} \nu_n(\{g\}) \int_{K_n} |f_\mu(gg') - h_n(gg')|\, d\omega_n(g') \\ &= \sum_{g \in T_n} \nu_n(\{g\}) \int_{K_n} |\gamma(g) f_\mu(g') - c_n^g|\, d\omega_n(g'),\end{aligned}$$

where c_n^g is the constant value of h_n on gK_n and γ is a (possibly discontinuous) character on G which agrees with f on the discrete measures in $M(G)$. It is apparent that the above expression will be minimized only if the c_n^g's are chosen to be of the form $c_n^g = c_n\gamma(g)$ for a constant c_n.

Now the sequence $\{h_n\}$ obviously converges to f_μ in $L^1(\mu)$-norm. Hence, $\int |h_n - h_{n-1}|\, d\mu \to 0$. Furthermore,

$$\begin{aligned}\int |h_n - h_{n-1}|\, d\mu &= \int |h_n - h_{n-1}|\, d\omega_n * \mu_n * \nu_{n-1}\\ &= \sum_{g\in T_{n-1}} \nu_{n-1}(\{g\}) \sum_{g'\in S_n} \mu_n(\{g'\}) \int_{K_n} |h_n(gg'g'') - h_{n-1}(gg'g'')|\, d\omega_n(g'')\\ &= \sum_{g\in T_{n-1}} \nu_{n-1}(\{g\}) \sum_{g'\in S_n} \mu_n(\{g'\})\, |\gamma(g)\gamma(g')c_n - \gamma(g)c_{n-1}|\\ &= \sum_{g'\in S_n} \mu_n(\{g'\})\, |\gamma(g')c_n - c_{n-1}|\\ &= \int |c_n\gamma - c_{n-1}|\, d\mu_n.\end{aligned}$$

Since $|h_n - h_m| \geqslant ||c_n| - |c_m||$, it is apparent that the sequence $\{|c_n|\}$ converges. If the limit is zero then $f_\mu = 0$. If the limit is a nonzero number r, we may as well assume the c_n's have constant modulus r. We then set $b_n = c_{n-1}c_n^{-1}$ and note that $|b_n| = 1$ and

$$\lim_{n\to\infty} \int |\gamma - b_n|\, d\mu_n = 0.$$

This leads to the following result of Johnson [3].

PROPOSITION. *Let $S = \bigcup_n S_n$ be the union of the supports of the measures μ_n. Suppose that whenever γ is a character of G and $\{b_n\}$ a sequence of constants of modulus one for which $\lim_{n\to\infty} \int |\gamma - b_n|\, d\mu_n = 0$ then γ agrees on S with a continuous character of G. Then for each $f \in \hat{S}$, $f_\mu = c\gamma$ for some c in the unit disc and some $\gamma \in \hat{G}$.*

PROOF. The hypothesis of the proposition implies that either $f_\mu = 0$ or we may assume that the character γ, for which $h_n(gg') = c_n\gamma(g)$ for $g \in T_n$ and $g' \in K_n$, is continuous. It follows that if $\{c_{n_j}\}$ is a subsequence of $\{c_n\}$ converging, say, to c, then $h_{n_j} \to c\gamma$ uniformly on a set supporting μ. Since $h_n \to f_\mu$ in $L^1(\mu)$, the proof is complete.

10.5.4. THE SHILOV BOUNDARY. With the above result, it is quite easy to construct infinite products $\mu = \Pi_{i=1}^\infty \mu_i$, of normalized positive discrete measures, for which μ satisfies (a) of 10.5.2. By 10.4.6, (b) of 10.5.2 will be satisfied if no power of μ is in a group algebra. In order to have (c) of 10.5.2 satisfied, it suffices to construct μ so that its Fourier transform $\hat{\mu}$ has modulus uniformly less than one off some compact subset of $\hat{G}$. It is quite easy to control $\hat{\mu}$ since it is simply $\Pi_{i=1}^\infty \hat{\mu}_i$.

The first construction of a measure satsifying all three properties was carried out on the group D_2 by Taylor [2]. Johnson [3] showed that such a construction was possible on any nondiscrete l. c. a. group. More recently, Moran [1] has shown that on any l. c. a. group G there is a product measure satisfying (a) and (b) for which $\hat{\mu}$ vanishes at infinity (hence, (c) is trivially satisfied). This shows that $M_0(G)$, as well as $M(G)$, has a proper Shilov boundary.

The propositions up to this point in §10.4 and §10.5 comprise the interesting part of the lore surrounding the Shilov boundary problem for $M(G)$. The actual construction for each G of a measure to which these results apply is a tedious and uninformative process. Hence, here we shall present only one example of such a construction. This is due to Johnson and takes care of the case $G = R$, which is perhaps the most difficult case.

10.5.5. AN EXAMPLE. For each i, let μ_i be the discrete measure on R with mass $2/3$ at 0 and $1/3$ at 2^{-i}. Note that each μ_i is positive and normalized. Furthermore by 10.4.1 the product measure $\omega_n = \Pi_{i=n+1}^{\infty} \mu_i$ is concentrated on $K_n = [0, 2^{-n})$. The support S_n of μ_n is $\{0, 2^{-n}\}$ and $K_n \cap (K_n + 2^{-n}) = \varnothing$. Hence, the measure $\mu = \Pi_{i=1}^{\infty} \mu_i$ is constructed as in 10.5.3.

Now if γ is any character of R, and $\gamma(2^{-n}) = \lambda_n$, then $\lambda_n = \lambda_{n+1}^2$ for each n. Furthermore,

$$\int |\gamma - b_n| \, d\mu_n = 2/3 \, |1 - b_n| + 1/3 \, |\lambda_n - b_n|.$$

Hence, if $\{b_n\}$ is chosen so that this expression converges to zero, we must have $b_n \to 1$ and $\lambda_n \to 1$. Thus, for large n, λ_{n+1} must be the principal square root of λ_n. It follows that on $S = \bigcup_n S_n$, γ must have the form $\gamma(t) = e^{itx}$ for some $x \in R$.

Since the hypothesis of Proposition 10.5.3 is satisfied, we conclude that (a) of 10.4.2 holds for μ.

Now the Fourier transform of μ_n is

$$\hat{\mu}_n(x) = 2/3 + 1/3 \exp(i2^{-n}x).$$

Hence

$$\hat{\mu}(x) = \prod_{n=1}^{\infty} (2/3 + 1/3 \exp(i2^{-n}x)).$$

Note that if $\pi 2^{n-1} \leqslant |x| \leqslant \pi 2^n$ then $\pi/2 \leqslant |2^{-n}x| \leqslant \pi$ and hence

$$|\hat{\mu}(x)| \leqslant |2/3 + 1/3 \exp(i\pi/2)| = \sqrt{5}/3.$$

We conclude that $|\hat{\mu}(x)| \leqslant \sqrt{5}/3 < 1$ off the set $(-\pi, \pi)$. It follows that (c) of 10.5.2 is satisfied by μ.

Recall that (b) of 10.5.2 will be satisfied if no power of μ is in a group algebra. However, since μ is certainly not discrete, this will follow if we can show that no power of μ is in $L(R)$.

If a power of μ were in $L(R)$ then certainly $\hat{\mu}$ would have to vanish at infinity. That it does not is easily seen by observing that $\hat{\mu}$ has the constant nonzero value

$$\prod_{n=1}^{\infty} (2/3 + 1/3 \exp(2\pi i 2^{-n}))$$

on the set of numbers of the form $2\pi 2^k$ $(k \geqslant 1)$.

Thus, we have established that μ satisfies the conditions of 10.5.2 and hence that $M(R)$ has a proper Shilov boundary.

REFERENCES

Amemiya, I. and Itô, T.

[1] *A simple proof of the theorem of P. J. Cohen,* Bull. Amer. Math. Soc. **70** (1964), 774–776. MR **29** #4862.

Arens, R.

[1] *The group of invertible elements of a commutative Banach algebra,* Studia Math. (1963), 21–23. MR **26** #4198. (Ser. Specjalna) Zeszyt **1**.

Arens, R. F. and Calderón, A. P.

[1] *Analytic functions of several Banach algebra elements,* Ann. of Math. (2) **62** (1955), 204–216. MR **17**, 177.

Arens, R. F. and Singer, I. M.

[1] *Generalized analytic functions,* Trans. Amer. Math. Soc. **81** (1956), 379–393. MR **17**, 1226.

Baartz, A.

[1] *The measure algebra of a locally compact semigroup,* Pacific J. Math. **21** (1967), 199–214. MR **35** #4678.

Baker, J. W.

[1] *Convolution measure algebras with involution,* Proc. Amer. Math. Soc. **27** (1971), 91–96. MR **43** #822.

Beurling, A.

[1] *Sur les intégrales de Fourier absolument convergentes et leur application a une transformation fonctionelle,* Nionde Skand. Mat.-kongres. (1938), Mercator, Helsingfors, 1939, pp. 345–366.

Bohr, H.

[1] *Über fastperiodische ebene bewegungen,* Comment. Math. Helv. **4** (1934), 51–64.

Bredon, G. E.

[1] *Sheaf theory,* McGraw-Hill, New York, 1967. MR **36** # 4552.

Brown, G.

[1] *On convolution measure algebras,* Proc. London Math. Soc. (3) **20** (1970), 490–506.

Brown, G. and Moran, W.

[1] *Idempotents and monothetic semigroups,* Proc. London Math. Soc. (3) **22** (1970), 203–216.

[2] *On the Shilov boundary of a measure algebra,* Bull. London Math. Soc. **3** (1971), 197–203.

[3] *A Dichotomy for infinite convolutions of discrete measures,* Proc. Cambridge Philos. Soc. (to appear).

Cohen, P.

[1] *On a conjecture of Littlewood and idempotent measures,* Amer. J. Math. **82** (1960), 191–212. MR **24** #A3231.

[2] *On homomorphisms of group algebras,* Amer. J. Math. **82** (1960), 213–226. MR **24** #A3232.

Douglas, R. and Taylor, J.

[1] *Wiener-Hopf operators with measure kernel.* Colloquia Math. Soc. János Bolyai 5. Hilbert space operators, Tihany (Hungary), 1970.

Dunkel, C. and Ramirez, D.

[1] *Topics in harmonic analysis,* Appleton-Century-Crofts, New York, 1971.

Edwards, R. E.

[1] *Fourier series: A modern introduction.* Vol. II, Holt-Rinehart-Winston, New York, 1967. MR **36** #5588.

Fisher, M. J.

[1] *The embeddability of an invertible measure.* Semigroup Forum (to appear).

Gamelin, T.

[1] *Uniform algebras,* Prentice-Hall, Englewood Cliffs, N. J., 1969.

Garnett, J.

[1] *A topological characterization of Gleason parts,* Pacific J. Math. **20** (1967), 59–63. MR **34** #4942.

Gleason, A. M.

[1] *Function algebras,* Seminars on Analytic Functions, II, Inst. for Advanced Study, Princeton, N. J., 1957.

[2] *The abstract theorem of Cauchy-Weil,* Pacific J. Math. **12** (1962), 511–525. MR **26** #5186.

Grothendieck, A.

[1] *Produits tensoriels topologiques et espaces nucléaires,* Mem. Amer. Math. Soc. No. 16 (1955). MR **17**, 763.

Gunning, R. C. and Rossi, H.

[1] *Analytic functions of several complex variables,* Prentice-Hall Series in Modern Analysis, Prentice-Hall, Englewood Cliffs, N. J., 1965. MR **31** #4927.

Hewitt, E.

[1] *The asymmetry of certain algebras of Fourier-Stieltjes transforms,* Michigan Math. J. **5** (1958), 149–158. MR **21** #4993.

[2] *Measure algebras on locally compact groups: a case history in functional analysis,* Studia Math. (Ser. Specjalna) Zeszyt **1** (1963), 41–52. MR **26** #5391.

Hewitt, E. and Kakutani, S.

[1] *A class of multiplicative linear functionals on the measure algebra of a locally compact Abelian group,* Illinois J. Math. **4** (1960), 553–574. MR **23** #A527.

[2] *Some multiplicative linear functionals on* $M(G)$, Ann. of Math. (2) **79** (1964), 489–505. MR **28** #4384.

Hewitt, E. and Ross, K.

[1] *Abstract harmonic analysis.* Vol. I: *Structure of topological groups. Integration theory, group representations,* Die Grundlehren der math. Wissenschaften, Bd. 115, Academic Press, New York; Springer-Verlag, Berlin, 1963. MR **28** #158.

Hewitt, E. and Zucherman, H.

[1] *Finite dimensional convolution algebras,* Acta. Math. **93** (1955), 67–119. MR **17**, 1048.

[2] *Harmonic analysis for certain semigroups,* Proc. Nat. Acad. Sci. U. S. A. **42** (1956), 253–255. MR **18**, 217.

[3] *Structure theory for a class of convolution algebras,* Pacific J. Math. **7** (1957), 913–941. MR **19**, 435.

[4] *Singular measures with absolutely continuous convolution squares,* Proc. Cambridge Philos. Soc. **62** (1966), 399–420. MR **33** #1655.

Hewitt, E. and Williamson, J. H.

[1] *Note on absolutely convergent Dirichlet series,* Proc. Amer. Math. Soc. **8** (1957), 863–868. MR **19**, 851.

Hoffman, K.

[1] *Analytic functions and logmodular Banach algebras,* Acta Math. **108** (1962), 271–317. MR **26** #6820.

Hofmann, K. H.

[1] *Categories with convergence, exponential functors, and the cohomology of compact abelian groups,* Math. Z. **104** (1968), 106–144. MR **37** #4195.

Hofmann, K. H. and Mostert, P.

[1] *Elements of compact semigroups,* Merrill, Columbus, Ohio, 1966. MR **35** #285.

Johnson, B.

[1] *Isometric isomorphisms of measure algebras,* Proc. Amer. Math. Soc. **15** (1964), 186–188. MR **28** #4056.

[2] *Symmetric maximal ideals in* $M(G)$, Proc. Amer. Math. Soc. **18** (1967), 1040–1044. MR **36** #1923.

[3] *The Šilov boundary of* $M(G)$, Trans. Amer. Math. Soc. **134** (1968), 289–296. MR **37** #5613.

Kakutani, S.

[1] *Concrete representation of abstract* (L)*-spaces and the mean ergodic theorem,* Ann. of Math. (2) **42** (1941), 523–537. MR **2**, 318.

[2] *Concrete representation of abstract (M)-spaces*, Ann. of Math. (2) **42** (1941), 994–1024. MR **3**, 7; MR **8**, 205.

Miller, R. R.

[1] *Gleason parts and Choquet boundary points in convolution measure algebras*, Pacific J. Math. **31** (1969), 755–771. MR **41** #2410.

Moran, W.

[1] *The Shilov boundary of $M_0(G)$*, Trans. Amer. Math. Soc. (to appear).

Naĭmark, M. A.

[1] *Normed rings*, GITTL, Moscow, 1956; English transl., Noordhoff, Groningen, 1959. MR **19**, 870; MR **22** #1824.

Newman, S. E.

[1] *Measure algebras on idempotent semigroups*, Pacific J. Math. **31** (1969), 161–169. MR **43** #945.

[2] *Measure algebras and functions of bounded variation on idempotent semigroups*, Bull. Amer. Math. Soc. **75** (1969), 1396–1400. MR **40** #4778.

Ramirez, D. E.

[1] *The measure algebra as an operator algebra*, Canad. J. Math. **20** (1968), 1391–1396. MR **38** #5010.

Rennison, J. F.

[1] *Arens products and measure algebras*, J. London Math. Soc. **44** (1969), 369–377. MR **38** #6367.

Rieffel, M. A.

[1] *A characterization of commutative group algebras and measure algebras*, Trans. Amer. Math. Soc. **116** (1965), 32–65. MR **33** #6300.

Ross, K.

[1] *The structure of certain measure algebras*, Pacific J. Math. **11** (1961), 723–737. MR **25** #463.

Royden, H. L.

[1] *Function algebras*, Bull. Amer. Math. Soc. **69** (1963), 281–298. MR **26** #6817.

Rudin, W.

[1] *Independent perfect sets in groups*, Michigan Math. J. **5** (1958), 159–161. MR **21** #4994.

[2] *Idempotent measures on Abelian groups*, Pacific J. Math. **9** (1959), 195–209. MR **21** #4332.

[3] *Measure algebras on abelian groups*, Bull. Amer. Math. Soc. **65** (1959), 227–247. MR **21** #7404.

[4] *Fourier-Stieltjes transforms of measures on independent sets*, Bull. Amer. Math. Soc. **66** (1960), 199–202. MR **22** #9802.

[5] *Fourier analysis on groups*, Interscience Tracts in Pure and Appl. Math., no. 12, Interscience, New York, 1962. MR **27** #2808.

Schaeffer, H. H.

[1] *Topological vector spaces,* MacMillan, New York, 1966. MR **33** #1689.

Schatten, R.

[1] *A theory of cross-spaces,* Ann. of Math. Studies, no. 26, Princeton Univ. Press, Princeton, N. J., 1950. MR **12**, 186.

Šilov, G. E.

[1] *On decomposition of a commutative normed ring in a direct sum of ideals,* Mat. Sb. **32 (74)** (1953), 353–364; English transl., Amer. Math. Soc. Transl. (2) **1** (1955), 37–48. MR **14**, 884; MR **17**, 512.

Simon, A. B.

[1] *Symmetry in measure algebras,* Bull. Amer. Math. Soc. **66** (1960), 399–400. MR **22** #9809.

Spanier, E. H.

[1] *Algebraic topology,* McGraw-Hill, New York, 1966. MR **35** #1007.

Šreider, Ju. A.

[1] *The structure of maximal ideals in rings of measures with convolution,* Mat. Sb. **27 (69)** (1950), 297–318; English transl., Amer. Math. Soc. Transl. (1) **8** (1962), 365–391. MR **12**, 420.

[2] *On an example of a generalized character,* Mat. Sb. **29 (71)** (1951), 419–426. (Russian) MR **13**, 755.

Taylor, J. L.

[1] *The structure of convolution measure algebras,* Trans. Amer. Math. Soc. **119** (1965), 150–166. MR **32** #2932.

[2] *The Shilov boundary of the algebra of measures on a group,* Proc. Amer. Math. Soc. **16** (1965), 941–945. MR **32** #2931.

[3] *Convolution measure algebras with group maximal ideal spaces,* Trans. Amer. Math. Soc. **128** (1967), 257–263. MR **35** #3375.

[4] *Ideal theory and Laplace transforms for a class of measure algebras on a group,* Acta Math. **121** (1968), 251–292. MR **38** #4923.

[5] *L-subalgebras of* $M(G)$, Trans. Amer. Math. Soc. **135** (1969), 105–113. MR **38** #1472.

[6] *Noncommutative convolution measure algebras,* Pacific J. Math. **31** (1969), 809–826. MR **41** #844.

[7] *Measures which are convolution exponentials,* Bull. Amer. Math. Soc. **76** (1970), 415–418. MR **41** #5986.

[8] *The cohomology of the spectrum of a measure algebra,* Acta Math. **126** (1971), 195–225.

[9] *On the spectrum of a measure,* Advances in Math. (to appear).

[10] *Inverses, logarithms, and idempotents in* $M(G)$, Rocky Mountain J. Math. **2** (1972), 183–206.

Varopoulos, N. Th.

[1] *Studies in harmonic analysis,* Proc. Cambridge Philos. Soc. **60** (1964), 465–516. MR **29** #1284.

[2] *Measure algebras of a locally compact group,* Séminaire Bourbaki, 1964/65 Exposé 282, Benjamin, New York, 1966. MR **33** #5420L.

Wallace, A. D.

[1] *The structure of topological semigroups,* Bull. Amer. Math. Soc. **61** (1955), 95–112. MR **16**, 796.

Wendel, J. G.

[1] *Left centralizers and isomophism of group algebras,* Pacific J. Math. **2** (1952), 251–261. MR **14**, 246.

Wiener, N.

[1] *Tauberian theorems,* Ann. of Math. **33** (1932), 1–100.

Wiener, N. and Pitt, R.

[1] *On absolutely convergent Fourier-Stieltjes transforms,* Duke Math. J. **4** (1938), 420–436.

Williamson, J. H.

[1] *A theorem on algebras of measures on topological groups,* Proc. Edinburgh Math. Soc. **11** (1958/59), 195–206. MR **22** #2851.